I0044551

Principles of Topography

Principles of Topography

Anna Nelson

SYRAWOOD
PUBLISHING HOUSE

New York

Published by Syrawood Publishing House,
750 Third Avenue, 9th Floor,
New York, NY 10017, USA
www.syrawoodpublishinghouse.com

Principles of Topography
Anna Nelson

© 2022 Syrawood Publishing House

International Standard Book Number: 978-1-64740-113-9 (Hardback)

This book contains information obtained from authentic and highly regarded sources. All chapters are published with permission under the Creative Commons Attribution Share Alike License or equivalent. A wide variety of references are listed. Permissions and sources are indicated; for detailed attributions, please refer to the permissions page. Reasonable efforts have been made to publish reliable data and information, but the authors, editors and publisher cannot assume any responsibility for the validity of all materials or the consequences of their use.

Trademark Notice: Registered trademark of products or corporate names are used only for explanation and identification without intent to infringe.

Cataloging-in-Publication Data

Principles of topography / Anna Nelson.
    p. cm.
Includes bibliographical references and index.
ISBN 978-1-64740-113-9
1. Physical geography. 2. Topographical surveying. 3. Geomorphology. 4. Earth (Planet)--Surface.
5. Topographic maps. I. Nelson, Anna.
GB60 .P75 2022
910.02--dc23

Table of Contents

Preface

This book has been written, keeping in view that students want more practical information. Thus, my aim has been to make it as comprehensive as possible for the readers. I would like to extend my thanks to my family and co-workers for their knowledge, support and encouragement all along.

Topography is an interdisciplinary field which focuses on the surface shape, features and description of land surfaces. It includes the records of the relief or terrain, identification of specific landforms, and three-dimensional quality of the surface. A topographic study is done for geological exploration and military planning. The objective is to determine the position of any feature and point in a horizontal coordinate system. There are numerous forms of topographic data which involves raw survey data, remote sensing data, etc. A topographic map is a type of map in modern mapping which is characterized by large scale detail. It is a quantitative representation of relief using contour lines. Various techniques used in topography include surveying, aerial photography and satellite imagery, photogrammetry, remote sensing, etc. This book aims to shed light on some of the unexplored aspects of topography. Different approaches, evaluations and methodologies on topography have been included in it. The topics covered in this book offer the readers new insights into this field.

A brief description of the chapters is provided below for further understanding:

Chapter – What is Topography?

Topography refers to the study of shapes and different characteristics of land surfaces such as hydrology, soil profile, vegetation, etc. It includes various principles of geoscience and planetary science. This chapter closely examines these concepts of topography to provide an extensive understanding of the subject.

Chapter – Types of Landform

A landform can be defined as natural or unnatural feature of the solid Earth surfaces. There are different types of landforms such as mountains, valleys, plains, plateaus, glaciers, canyon, volcanoes, bay, mid-ocean ridge, etc. The topics elaborated in this chapter will help in gaining a better perspective about different types of topography.

Chapter – Different Approaches to Study Topography

There are a number of ways in which topography can be studied. These include aerial survey, land surveying, compass surveying, remote sensing, photogrammetry, satellite imagery, digital elevation model, geovisualization, hypsometric tinting, etc. These diverse approaches for studying topography in the current scenario have been thoroughly discussed in this chapter.

Chapter – Cartography

The science of making maps and charts is referred to as cartography. It helps in communicating spatial, topographic and geographic information of any landform, and terrain. Map projection, cartographic labeling, cartographic generalization, etc. are some of its principles. All these diverse principles of cartography have been carefully analyzed in this chapter.

Chapter – Geographic Information System

Geographic information refers to the data associated with any specific location. A system which collects, stores, analysis and manages such data is termed as geographic information system. Some of its processes are spatial analysis, participatory geographic information system, CyberGIS, etc. This chapter discusses in detail these processes related to geographic information system.

Chapter – Allied Fields

Topography is an interdisciplinary subject plays vital roles in many other fields. Some of them are earth sciences, planetary science, geography, geology, geomorphology, geoscience, etc. This chapter has been carefully written to provide an in-depth understanding of the varied facets of these allied fields of topography.

Anna Nelson

1

What is Topography?

Topography refers to the study of shapes and different characteristics of land surfaces such as hydrology, soil profile, vegetation, etc. It includes various principles of geoscience and planetary science. This chapter closely examines these concepts of topography to provide an extensive understanding of the subject.

Topography is the study of Earth's surface features or those of planets, moons, and asteroids.

In the broadest sense, topography is concerned with local detail in general, including not only relief but also vegetative and human-made features, and even local history and culture. This meaning is less common in America, where topographic maps with elevation contours have made "topography" synonymous with relief. The older sense of topography as the study of place still has currency in Europe.

Here topography specifically involves the recording of relief or terrain, the three-dimensional quality of the surface, and the identification of specific landforms. In modern usage, this involves generation of elevation data in electronic form. It is often considered to include the graphic representation of the landform on a map by a variety of techniques, including contour lines, Hypsometric tints, and relief shading.

Objectives

The objective of topography is to determine the position of any feature or more generally any point in terms of both a horizontal coordinate system such as latitude and longitude, and elevation. Identifying (naming) features and recognizing typical landform patterns are also part of the field. Equally important is the accurate portrayal of multidimensional features, naturally occurring or human made, in two-dimensional form. Prior to the development of remote sensing techniques, unusual techniques were used. Wooden carvings were used by native peoples to depict coastlines and elevations; "stick charts" were created by Marshall Islanders to not only map distances, but also to note currents and wave fronts.

A topographic study may be made for a variety of reasons: military planning and geological exploration have been primary motivators to start survey programs, but detailed information about terrain and surface features is essential for the planning and construction of any major civil engineering, public works, or reclamation projects. Environmental research, travel, and sports are all beneficiaries of accurate representations of terrain and relief.

Techniques of Topography

There are a variety of approaches to studying topography. The most appropriate method to use depends on the scale and size of the area under study, its accessibility, and the quality of existing surveys.

Direct Survey

A surveying point.

Surveying helps determine accurately the terrestrial or three-dimensional position of points and the distances and angles between them using surveying instruments.

Even though remote sensing has greatly speeded up the process of gathering information, and has allowed greater accuracy control over long distances, the direct survey still provides the basic control points and framework for all topographic work, whether manual or Geographic Information Systems (GIS)- based.

In areas where there has been an extensive direct survey and mapping program (most of Europe and the Continental US, for example), the compiled data forms the basis of basic digital elevation datasets such as United States Geological Survey Digital Elevation Model data. This data must often be edited to eliminate discrepancies between surveys, but still forms a valuable set of information for large-scale analysis.

The original American topographic surveys (or the British "Ordnance" surveys) involved not only recording of relief, but identification of landmark features and vegetative land cover.

Remote Sensing

In the broadest sense, remote sensing is the short or large-scale acquisition of information of an object or phenomenon, by the use of either recording or real-time sensing device(s) that is not in physical or intimate contact with the object (such as by way of aircraft, spacecraft, satellite, buoy, or ship). Methods of remote sensing include:

Aerial and Satellite Imagery

Besides their role in photogrammetry, aerial and satellite imagery can be used to identify and delineate terrain features and more general land-cover features. These types of images increasingly have become part of geovisualization, whether as maps or GIS depictions. False-color and non-visible spectra imaging can also help determine the lie of the land by delineating vegetation and other land-use information more clearly. Images can be in visible colors and in other spectra.

Synthetic aperture radar image of Death
Valley colored using polarimetry.

Photogrammetry

Photogrammetry is a measurement technique for which the coordinates of the points of a multi-dimensional object are determined by measurements made in two photographic images (or more) taken starting from different positions, usually from different passes of an aerial photography flight. In this technique, the common points are identified on each image. A line of sight (or ray) can be derived from the camera location to the point on the object. The intersection of these rays (triangulation) determines the relative three-dimensional position of the point. Known control points can be used to give these relative positions absolute values. More sophisticated algorithms can exploit other information on the scene already known.

Radar and Sonar

Satellite radar mapping is one of the major techniques of generating Digital Elevation Models. Seismographic information can be useful in mapping sub-surface structures. Similar techniques are applied in bathymetric surveys using sonar or depth soundings to determine the terrain of the ocean floor.

Forms of Topographic Data

Terrain is typically modeled using either vector (Triangulated Irregular Network or TIN) or gridded (Raster image) mathematical models. In most uses in environmental sciences, land surface is represented and modeled using gridded models. In civil engineering, for example, most representations of land surface employ some variant of TIN models. In geostatistics, land surface is commonly modeled as a combination of the two signals—the smooth (spatially correlated) and the rough (noise) signal.

In practice, surveyors first sample heights in an area, then use these to produce a Digital Land Surface Model (also known as a digital elevation model). The DLSM can then be used to visualize terrain, drape remote sensing images, quantify ecological properties of a surface or extract land surface objects. Note that the contour data or any other sampled elevation datasets are not a DLSM. A DLSM implies that elevation is available continuously at each location in the study area, i.e. that the map represents a complete surface. Digital Land Surface Models should not be confused with Digital Surface Models, which can be surfaces of the canopy, buildings and similar objects. For example, in the case of surface models produces using the LIDAR technology, one can have several surfaces—starting from the top of the canopy to the actual solid earth. The difference between the two surface models can then be used to derive volumetric measures (height of trees etc).

Raw Survey Data

Topographic survey information is historically based upon the notes of surveyors who may have derived naming and cultural information from other local sources (for example, boundary delineation may be derived from local cadastral mapping). While of historical interest, these field notes inevitably include errors and contradictions that later stages in map production resolve.

Remote Sensing Data

As with field notes, remote sensing data (aerial and satellite photography, for example), is raw and uninterpreted. It may contain gaps (due to cloud cover for example) or inconsistencies (due to the timing of specific image captures). Most modern topographic mapping includes a large component of remotely sensed data in its compilation process.

Topographic Mapping

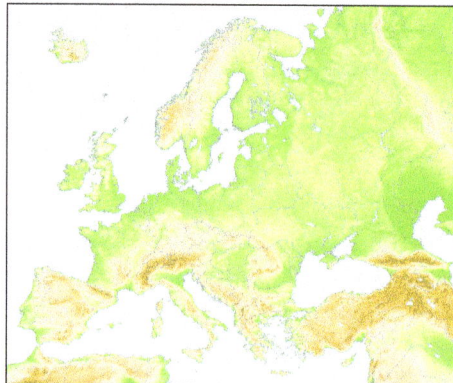

A map of Europe using elevation modeling.

In its contemporary definition, topographic mapping shows relief. In the United States, USGS topographic maps show relief using contour lines. The USGS calls maps based on topographic surveys, but without contours, "planimetric maps."

These planimetric maps show not only the contours, but also any significant streams or other bodies of water, forest cover, built-up areas or individual buildings (depending on scale), and other features and points of interest.

While not officially "topographic" maps, the national surveys of other nations share many of the same features, and so they are often generally called "topographic maps."

Existing topographic survey maps, because of their comprehensive and encyclopedic coverage, form the basis for much derived topographic work, thematic maps, for example. Digital Elevation Models, for example, have often been created not from new remote sensing data but from existing paper topographic maps. Many government and private publishers use the artwork (especially the contour lines) from existing topographic map sheets as the basis for their own specialized or updated topographic maps.

Topographic mapping should not be confused with Geologic mapping. The latter is concerned with underlying structures and processes beneath the surface, rather than with identifiable surface features.

Digital Elevation Modelling

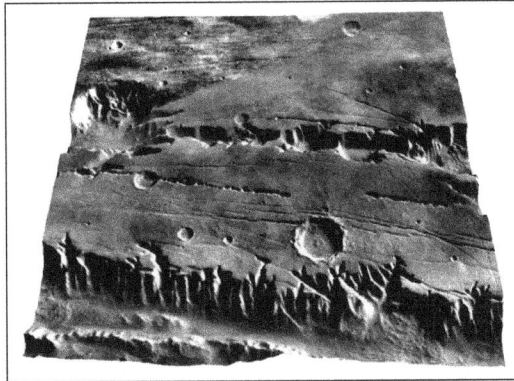

3D rendering of a DEM used for the topography of Mars.

The digital elevation model (DEM) is a raster-based digital dataset of the topography (altimetry and/or bathymetry) of all or part of the Earth (or a telluric planet). The pixels of the dataset are each assigned an elevation value, and a header portion of the dataset defines the area of coverage, the units each pixel covers, and the units of elevation (and the zero-point). DEMs may be derived from existing paper maps and survey data, or they may be generated from new satellite or other remotely-sensed radar or sonar data.

Topological Modelling

A geographic information system (GIS) can recognize and analyze the spatial relationships that exist within digitally stored spatial data. These topological relationships allow complex spatial modeling and analysis to be performed. Topological relationships between geometric entities traditionally include adjacency (what adjoins what), containment (what encloses what), and proximity (how close something is to something else). These are used to:

- Reconstitute a sight in synthesized images of the ground,

- Determine a trajectory of overflight of the ground,

- Calculate surfaces or volumes,

- Trace topographic profiles,

- Handle in a quantitative way the studied ground.

Topography in other Fields

Topography has been applied to different science fields. In neuroscience, the neuroimaging discipline uses techniques such as EEG topography for brain mapping. In ophthalmology, corneal topography is used as a technique for mapping the surface curvature of the cornea.

2
Types of Landform

A landform can be defined as natural or unnatural feature of the solid Earth surfaces. There are different types of landforms such as mountains, valleys, plains, plateaus, glaciers, canyon, volcanoes, bay, mid-ocean ridge, etc. The topics elaborated in this chapter will help in gaining a better perspective about different types of topography.

Continental Landform

A Continental landform is a natural or artificial feature of the solid surface of the Earth. Familiar examples are mountains (including volcanic cones), plateaus, and valleys. (The term landform also can be applied to related features that occur on the floor of the Earth's ocean basins, as, for example, seamounts, mid-oceanic ridges, and submarine canyons). Such structures are rendered unique by the tectonic mechanisms that generate them and by the climatically controlled denudational systems that modify them through time. The resulting topographic features tend to reflect both the tectonic and the denudational processes involved.

The most dramatic expression of tectonism is mountainous topography, which is either generated along continental margins by collisions between the slablike plates that make up the Earth's lithosphere or formed somewhat farther inland by rifting and faulting. Far more subtle tectonic expressions are manifested by the vast continental regions of limited relief and elevation affected by gentle uplift, subsidence, tilting, and warping. The denudational processes act upon the tectonic "stage set" and are able to modify its features in a degree that reflects which forces are dominant through time. Volcanism as a syn-tectonic phenomenon may modify any landscape by fissure-erupted flood basalts capable of creating regional lava plateaus or by vent eruptions that yield individual volcanoes.

The denudational processes, which involve rock weathering and both erosion and deposition of rock debris, are governed in character by climate, whose variations of heat and moisture create vegetated, desert, or glacial expressions. Most regions have been exposed to repeated changes in climate rather than to a single enduring condition. Climates can change very slowly through continental drift and much more rapidly through variations in such factors as solar radiation.

In most instances, a combination of the foregoing factors is responsible for any given landscape. In a few cases, tectonism, some special combination of denudational effects, or volcanism may control the entire landform suite. Where tectonism exists in the form of orogenic uplift, the high-elevation topography depends on the nature of denudation. In humid or glacial environments whose geo-morphic agencies can exploit lithologic variations, the rocks are etched into mountainous relief like that of the Alps or the southern Andes. In arid orogenic settings, the effects of aggradation and planation often result in alluviated intermontane basins that merge with high plateaus interrupted or bordered by mountains such as the central Andes or those of Tibet and Colorado in the western United States.

In continental regions where mountainous uplifts are lacking, denudational processes operate on rocks that are only slightly deformed—if they are sedimentary—and only moderately elevated. This produces broad basins, ramps, swells, and plains. These are most thoroughly dissected in rain-and-river environments (sometimes attaining local mountainous relief on uplifts). Elsewhere, they may be broadly alluviated and pedimented where mainly arid, or widely scoured and aggraded where glacial.

Minor denudational landforms are superimposed on the major features already noted. Where aridity has dominated, they include pediments, pans, dune complexes, dry washes, alluvial ve-neers, bajadas, and fans. Ridge-ravine topography and integrated drainage networks with associ-ated thick soils occur where humid conditions have prevailed. Combinations of these features are widespread wherever arid and humid conditions have alternated, and either category may merge laterally with the complex suite of erosional and depositional landforms generated by continental glaciers at higher latitudes.

Mountains

A mountain is a large landform that rises above the surrounding land in a limited area, usually in the form of a peak. A mountain is generally steeper than a hill. Mountains are formed through tec-tonic forces or volcanism. These forces can locally raise the surface of the earth. Mountains erode slowly through the action of rivers, weather conditions, and glaciers. A few mountains are isolated summits, but most occur in huge mountain ranges.

High elevations on mountains produce colder climates than at sea level. These colder climates strongly affect the ecosystems of mountains: different elevations have different plants and ani-mals. Because of the less hospitable terrain and climate, mountains tend to be used less for agri-culture and more for resource extraction and recreation, such as mountain climbing.

Mount Ararat, as seen from Armenia.

The highest mountain on Earth is Mount Everest in the Himalayas of Asia, whose summit is 8,850 m (29,035 ft) above mean sea level. The highest known mountain on any planet in the Solar System is Olympus Mons on Mars at 21,171 m (69,459 ft).

Peaks of Mount Kenya.

Mount Wilhelm in Papua New Guinea.

There is no universally accepted definition of a mountain. Elevation, volume, relief, steepness, spacing and continuity have been used as criteria for defining a mountain. In the Oxford English Dictionary a mountain is defined as "a natural elevation of the earth surface rising more or less abruptly from the surrounding level and attaining an altitude which, relatively to the adjacent elevation, is impressive or notable."

Whether a landform is called a mountain may depend on local usage. Mount Scott outside Lawton, Oklahoma, USA, is only 251 m (823 ft) from its base to its highest point. Whittow's Dictionary of Physical Geography states "Some authorities regard eminences above 600 metres (2,000 ft) as mountains, those below being referred to as hills."

In the United Kingdom and the Republic of Ireland, a mountain is usually defined as any summit at least 2,000 feet (610 m) high, which accords with the official UK government's definition that a mountain, for the purposes of access, is a summit of 2,000 feet (610 m) or higher. In addition, some definitions also include a topographical prominence requirement, typically 100 or 500 feet (30 or 152 m). At one time the U.S. Board on Geographic Names defined a mountain as being 1,000 feet (300 m) or taller, but has abandoned the definition since the 1970s. Any similar landform lower than this height was considered a hill. However, today, the United States Geological Survey (USGS) concludes that these terms do not have technical definitions in the US.

The UN Environmental Programme's definition of "mountainous environment" includes any of the following:

- Elevation of at least 2,500 m (8,200 ft);

- Elevation of at least 1,500 m (4,900 ft), with a slope greater than 2 degrees;

- Elevation of at least 1,000 m (3,300 ft), with a slope greater than 5 degrees;

- Elevation of at least 300 m (980 ft), with a 300 m (980 ft) elevation range within 7 km (4.3 mi).

Using these definitions, mountains cover 33% of Eurasia, 19% of South America, 24% of North America, and 14% of Africa. As a whole, 24% of the Earth's land mass is mountainous.

There are three main types of mountains: volcanic, fold, and block. All three types are formed from plate tectonics: when portions of the Earth's crust move, crumple, and dive. Compressional forces, isostatic uplift and intrusion of igneous matter forces surface rock upward, creating a landform higher than the surrounding features. The height of the feature makes it either a hill or, if higher and steeper, a mountain. Major mountains tend to occur in long linear arcs, indicating tectonic plate boundaries and activity.

Volcanoes

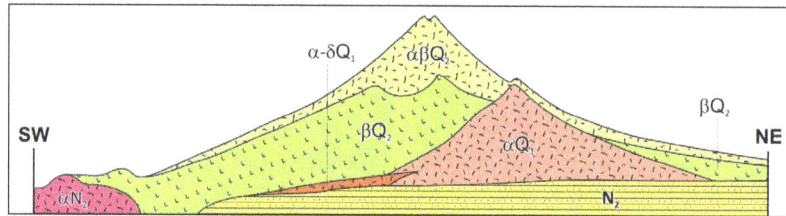
Geological cross-section of Fuji volcano.

Volcanoes are formed when a plate is pushed below another plate, or at a mid-ocean ridge or hotspot. At a depth of around 100 km, melting occurs in rock above the slab (due to the addition of water), and forms magma that reaches the surface. When the magma reaches the surface, it often builds a volcanic mountain, such as a shield volcano or a stratovolcano. Examples of volcanoes include Mount Fuji in Japan and Mount Pinatubo in the Philippines. The magma does not have to reach the surface in order to create a mountain: magma that solidifies below ground can still form dome mountains, such as Navajo Mountain in the US.

Fold Mountains

Illustration of mountains that developed on a fold that thrusted.

Fold mountains occur when two plates collide: shortening occurs along thrust faults and the crust is overthickened. Since the less dense continental crust "floats" on the denser mantle rocks beneath, the weight of any crustal material forced upward to form hills, plateaus or mountains must be balanced by the buoyancy force of a much greater volume forced downward into the mantle. Thus the continental crust is normally much thicker under mountains, compared to lower lying areas. Rock can fold either symmetrically or asymmetrically. The upfolds are anticlines and the downfolds are synclines: in asymmetric folding there may also be recumbent and overturned folds. The Balkan Mountains and the Jura Mountains are examples of fold mountains.

Block Mountains

Pirin Mountain, Bulgaria, part of the fault-block Rila-Rhodope massif.

Block mountains are caused by faults in the crust: a plane where rocks have moved past each other. When rocks on one side of a fault rise relative to the other, it can form a mountain. The uplifted blocks are block mountains or horsts. The intervening dropped blocks are termed graben: these can be small or form extensive rift valley systems. This form of landscape can be seen in East Africa, the Vosges, the Basin and Range Province of Western North America and the Rhine valley. These areas often occur when the regional stress is extensional and the crust is thinned.

Erosion

The Catskills in Upstate New York represent an eroded plateau.

During and following uplift, mountains are subjected to the agents of erosion (water, wind, ice, and gravity) which gradually wear the uplifted area down. Erosion causes the surface of mountains to be younger than the rocks that form the mountains themselves. Glacial processes produce characteristic landforms, such as pyramidal peaks, knife-edge arêtes, and bowl-shaped cirques that can contain lakes. Plateau mountains, such as the Catskills, are formed from the erosion of an uplifted plateau.

In earth science, erosion is the action of surface processes (such as water flow or wind) that removes soil, rock, or dissolved material from one location on the Earth's crust, and then transport it away to another location. The particulate breakdown of rock or soil into clastic sediment is referred to as physical or mechanical erosion; this contrasts with chemical erosion, where soil or rock material is removed from an area by its dissolving into a solvent (typically water), followed by the flow away of that solution. Eroded sediment or solutes may be transported just a few millimetres, or for thousands of kilometres.

Climate

A combination of high latitude and high altitude makes the northern
Urals in picture to have climatic conditions that make the ground barren.

Climate in the mountains becomes colder at high elevations, due to an interaction between radiation and convection. Sunlight in the visible spectrum hits the ground and heats it. The ground then heats the air at the surface. If radiation were the only way to transfer heat from the ground to space, the greenhouse effect of gases in the atmosphere would keep the ground at roughly 333 K (60 °C; 140 °F), and the temperature would decay exponentially with height.

However, when air is hot, it tends to expand, which lowers its density. Thus, hot air tends to rise and transfer heat upward. This is the process of convection. Convection comes to equilibrium when a parcel of air at a given altitude has the same density as its surroundings. Air is a poor conductor of heat, so a parcel of air will rise and fall without exchanging heat. This is known as an adiabatic process, which has a characteristic pressure-temperature dependence. As the pressure gets lower, the temperature decreases. The rate of decrease of temperature with elevation is known as the adiabatic lapse rate, which is approximately 9.8 °C per kilometre (or 5.4 °F per 1000 feet) of altitude.

Note that the presence of water in the atmosphere complicates the process of convection. Water vapor contains latent heat of vaporization. As air rises and cools, it eventually becomes saturated and cannot hold its quantity of water vapor. The water vapor condenses (forming clouds), and releases heat, which changes the lapse rate from the dry adiabatic lapse rate to the moist adiabatic lapse rate (5.5 °C per kilometre or 3 °F per 1000 feet). The actual lapse rate can vary by altitude and by location.

Therefore, moving up 100 metres on a mountain is roughly equivalent to moving 80 kilometres (45 miles or 0.75° of latitude) towards the nearest pole. This relationship is only approximate, however, since local factors such as proximity to oceans (such as the Arctic Ocean) can drastically modify the climate. As the altitude increases, the main form of precipitation becomes snow and the winds increase.

The effect of the climate on the ecology at an elevation can be largely captured through a combination of amount of precipitation, and the biotemperature, as described by Leslie Holdridge in 1947. Biotemperature is the mean temperature; all temperatures below 0 °C (32 °F) are considered to be 0 °C. When the temperature is below 0 °C, plants are dormant, so the exact temperature is unimportant. The peaks of mountains with permanent snow can have a biotemperature below 1.5 °C (34.7 °F).

Ecology

The colder climate on mountains affects the plants and animals residing on mountains. A particular set of plants and animals tend to be adapted to a relatively narrow range of climate. Thus, ecosystems tend to lie along elevation bands of roughly constant climate. This is called altitudinal zonation. In regions with dry climates, the tendency of mountains to have higher precipitation as well as lower temperatures also provides for varying conditions, which enhances zonation.

Some plants and animals found in altitudinal zones tend to become isolated since the conditions above and below a particular zone will be inhospitable and thus constrain their movements or dispersal. These isolated ecological systems are known as sky islands.

An alpine mire in the Swiss Alps.

Altitudinal zones tend to follow a typical pattern. At the highest elevations, trees cannot grow, and whatever life may be present will be of the alpine type, resembling tundra. Just below the tree line, one may find subalpine forests of needleleaf trees, which can withstand cold, dry conditions. Below that, montane forests grow. In the temperate portions of the earth, those forests tend to be needleleaf trees, while in the tropics, they can be broadleaf trees growing in a rain forest.

Valleys

A valley is a low area between hills or mountains typically with a river running through it. In geology, a valley or dale is a depression that is longer than it is wide. The terms U-shaped and V-shaped are descriptive terms of geography to characterize the form of valleys. Most valleys belong to one of these two main types or a mixture of them, at least with respect to the cross section of the slopes or hillsides.

River Valleys

A valley formed by flowing water, called *fluvial valley* or *river valley*, is usually V-shaped. The exact shape will depend on the characteristics of the stream flowing through it. Rivers with steep gradients, as in mountain ranges, produce steep walls and a bottom. Shallower slopes may produce broader and gentler valleys. However, in the lowest stretch of a river, where it approaches its base level, it begins to deposit sediment and the valley bottom becomes a floodplain.

Valley of Palakaria river springing from Vitosha Mountain,
seen in the background, in Bulgaria.

Some broad *V* examples are:

- North America: Black Canyon of the Gunnison National Park, and others in Grand Canyon NP.

- Europe:

    ◦ Austria: narrow passages of upper Inn valley (Inntal), affluents of Enns.

    ◦ Switzerland: Napf region, Zurich Oberland, Engadin.

    ◦ Germany: affluents to the middle reaches of Rhine and Mosel.

Some of the first human complex societies originated in river valleys, such as that of the Nile, Tigris-Euphrates, Indus, Ganges, Yangtze, Yellow River, Mississippi, and arguably Amazon. In prehistory, the rivers were used as a source of fresh water and food (fish and game), as well as a place to wash and a sewer. The proximity of water moderated temperature extremes and provided a source for irrigation, stimulating the development of agriculture. Most of the first civilizations developed from these river valley communities.

Vales

In geography, a *vale* is a wide river valley, usually with a particularly wide flood plain or flat valley bottom.

In Southern England, vales commonly occur between the escarpment slopes of pairs of chalk formations, where the chalk dome has been eroded, exposing less resistant underlying rock, usually claystone.

Rift Valleys

Rift valleys, such as the Albertine Rift and Gregory Rift are formed by the expansion of the Earth's crust due to tectonic activity beneath the Earth's surface.

Glacial Valleys

There are various forms of valley associated with glaciation that may be referred to as glacial valleys.

Yosemite Valley from an airplane.

U-shaped or Trough Valleys

U-shaped valley on the Afon Fathew near Dolgoch, Wales.

A valley carved by glaciers is normally U-shaped and resembles a trough. This trough valley becomes visible upon the recession of the glacier that forms it. When the ice recedes or thaws, the valley remains, often littered with small boulders that were transported within the ice. Floor gradient does not affect the valley's shape, it is the glacier's size that does. Continuously flowing glaciers – especially in the ice age – and large-sized glaciers carve wide, deep incised valleys, sometimes with valley steps that reflect differing erosion rates.

A panoramic view of two merging U-shaped
valleys in Pirin mountain, Bulgaria.

Examples of U-shaped valleys are found in every mountainous region that has experienced glaciation, usually during the Pleistocene ice ages. Most present U-shaped valleys started as V-shaped before glaciation. The glaciers carved it out wider and deeper, simultaneously changing the shape. This proceeds through the glacial erosion processes of glaciation and abrasion, which results in large rocky material (glacial till) being carried in the glacier. A material called boulder clay is deposited on the floor of the valley. As the ice melts and retreats, the valley is left with very steep sides and a wide, flat floor. A river or stream may remain in the valley. This replaces the original stream or river and is known as a misfit stream because it is smaller than one would expect given the size of its valley.

Other interesting glacially carved valleys include:

- Yosemite Valley (United States).

- Side valleys of the Austrian river Salzach for their parallel directions and hanging mouths.

- Some Scottish glens full with bushes and flowers.

- That of the St. Mary River in Glacier National Park in Montana, USA.

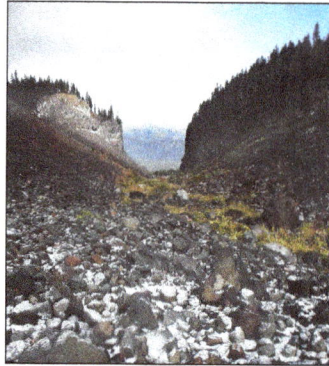

A glaciated valley in the Mount Hood Wilderness showing a characteristic
U-shape, the bottom's rocky 'rubble' accretion and the broad shoulders.

Tunnel Valleys

A tunnel valley is a large, long, U-shaped valley originally cut under the glacial ice near the margin of continental ice sheets such as that now covering Antarctica and formerly covering portions of all continents during past glacial ages.

A tunnel valley can be up to 100 km (62 mi), 4 km (2.5 mi) wide, and 400 m (1,300 ft) deep (its depth may vary along its length).

Tunnel valleys were formed by subglacial erosion by water. They served as subglacial drainage pathways carrying large volumes of melt water. Their cross-sections exhibit steep-sided flanks similar to fjord walls, and their flat bottoms are typical of subglacial glacial erosion.

Meltwater Valleys

In northern Central Europe, the Scandinavian ice sheet during the various ice ages advanced slightly uphill against the lie of the land. As a result, its meltwaters flowed parallel to the ice margin to reach the North Sea basin, forming huge, flat valleys known as *Urstromtäler*. Unlike the other forms of glacial valley, these were formed by glacial meltwaters.

New Zealand's Hooker Valley at Aoraki/Mount Cook National Park,
with Hooker Glacier's terminus at Hooker Lake in the background.

Transition Forms and Valley Shoulders

Depending on the topography, the rock types and the climate, a lot of transitional forms between V-, U- and plain valleys exist. Their bottoms can be broad or narrow, but characteristic is also the type of valley shoulder. The broader a mountain valley, the lower its shoulders are located in most cases. An important exception is canyons where the shoulder almost is near the top of the valley's slope. In the Alps – e.g. the Tyrolean Inn valley – the shoulders are quite low (100–200 meters above the bottom). Many villages are located here (esp. at the sunny side) because the climate is very mild: even in winter when the valley's floor is completely filled with fog, these villages are in sunshine.

Look from Paria View to a valley in Bryce Canyon, Utah, with very striking shoulders.

In some stress-tectonic regions of the Rockies or the Alps (e.g. Salzburg) the side valleys are parallel to each other, and additionally they are hanging. The brooks flow into the river in form of deep canyons or waterfalls. Usually this fact is the result of a violent erosion of the former valley shoulders, a special genesis found also at arêtes and glacial cirques, at every Scottish glen, or in a northern fjord.

Hanging Valleys

Bridal Veil Falls in Yosemite National
Park flowing from a hanging valley.

A hanging valley is a tributary valley that is higher than the main valley. They are most commonly associated with U-shaped valleys when a tributary glacier flows into a glacier of larger volume. The main glacier erodes a deep U-shaped valley with nearly vertical sides while the tributary glacier, with a smaller volume of ice, makes a shallower U-shaped valley. Since the surfaces of the glaciers were originally at the same elevation, the shallower valley appears to be 'hanging' above the main valley. Often, waterfalls form at or near the outlet of the upper valley.

Hanging valleys also occur in fjord systems under water. The branches of Sognefjord are for instance much shallower than the main fjord. The mouth of Fjærlandsfjord is about 400 meters deep while the main fjord is 1200 meters nearby. The mouth of Ikjefjord is only 50 meters deep while the main fjord is around 1300 meters at the same point.

Hanging valley, Ibar (river) valley, Rila Mountain, Bulgaria.

Glaciated terrain is not the only site of hanging streams and valleys. Hanging valleys are also simply the product of varying rates of erosion of the main valley and the tributary valleys. The varying rates of erosion are associated with the composition of the adjacent rocks in the different valley locations. The tributary valleys are eroded and deepened by glaciers or erosion at a slower rate than that of the main valley floor, thus the difference in the two valleys' depth increases over time. The tributary valley composed of more resistant rock then hangs over the main valley.

Trough-shaped Valleys

Trough-shaped valleys also form in regions of heavy denudation. By contrast, with glacial U-shaped valleys, there is less downward and sideways erosion. The severe slope denudation results in gently sloping valley sides and their transition to the actual valley bottom is unclear. Trough-shaped valleys occur mainly in periglacial regions and in tropical regions of variable wetness. Both climates are dominated by heavy denudation.

Box Valleys

Box valleys have wide, relatively level floors and steep sides. They are common in periglacial areas and occur in mid-latitudes, but also occur in tropical and arid regions.

Valley Floors

Usually the bottom of a main valley is broad – independent of the U or V shape. It typically ranges from about one to ten kilometers in width and is commonly filled with mountain sediments. The shape of the floor can be rather horizontal, similar to a flat cylinder, or terraced.

Side valleys are rather V than U-shaped; near the mouth waterfalls are possible if it is a hanging valley. The location of the villages depends on the across-valley profile, on climate and local traditions, and on the danger of avalanches or landslides. Predominant are places on terraces or alluvial fans if they exist.

Historic siting of villages within the mainstem valleys, however, have chiefly considered the potential of flooding.

Plains

A plain is a flat, sweeping landmass that generally does not change much in elevation. Plains occur as lowlands along the bottoms of valleys or on the doorsteps of mountains, as coastal plains, and as plateaus or uplands.

In a valley, a plain is enclosed on two sides, but in other cases a plain may be delineated by a complete or partial ring of hills, by mountains, or by cliffs. Where a geological region contains more than one plain, they may be connected by a pass (sometimes termed a gap). Coastal plains would mostly rise from sea level until they run into elevated features such as mountains or plateaus.

Plains are one of the major landforms on earth, where they are present on all continents, and would cover more than one-third of the world's land area. Plains may have been formed from flowing lava, deposited by water, ice, wind, or formed by erosion by these agents from hills and mountains. Plains would generally be under the grassland (temperate or subtropical), steppe (semi-arid), savannah (tropical) or tundra (polar) biomes. In a few instances, deserts and rainforests can also be plains.

Plains in many areas are important for agriculture because where the soils were deposited as sediments they may be deep and fertile, and the flatness facilitates mechanization of crop production; or because they support grasslands which provide good grazing for livestock.

Types of Plain

Depositional Plains

A small, incised alluvial plain from Red Rock Canyon State Park (California).

Depositional plains formed by the deposition of materials brought by various agents of transportation such as glaciers, rivers, waves, and wind. Their fertility and economic relevance depend greatly on the types of sediments that are laid down. The types of depositional plains include:

- Abyssal plains, flat or very gently sloping areas of the deep ocean basin.

- Planitia, the Latin word for plain, is used in the naming of plains on extraterrestrial objects (planets and moons), such as Hellas Planitia on Mars or Sedna Planitia on Venus.

- Alluvial plains, which are formed by rivers and which may be one of these overlapping types:

  ○ Alluvial plains, formed over a long period of time by a river depositing sediment on their flood plains or beds, which become alluvial soil. The difference between a flood plain and an alluvial plain is: a flood plain represents areas experiencing flooding fairly regularly in the present or recently, whereas an alluvial plain includes areas where a flood plain is now and used to be, or areas which only experience flooding a few times a century.

  ○ Flood plain, adjacent to a lake, river, stream, or wetland that experiences occasional or periodic flooding.

  ○ Scroll plain, a plain through which a river meanders with a very low gradient.

- Glacial plains, formed by the movement of glaciers under the force of gravity:

  ○ Outwash plain (also known as sandur; plural sandar), a glacial out-wash plain formed of sediments deposited by melt-water at the terminus of a glacier. Sandar consist mainly of stratified (layered and sorted) gravel and sand.

  ○ Till plains, plain of glacial till that form when a sheet of ice becomes detached from the main body of a glacier and melts in place depositing the sediments it carries. Till plains are composed of unsorted material (till) of all sizes.

- Lacustrine plains, plains that originally formed in a lacustrine environment, that is, as the bed of a lake.

- Lava plains, formed by sheets of flowing lava.

A flood plain in the Isle of Wight.

Erosional Plains

Erosional plains have been leveled by various agents of denudation such as running water, rivers, wind and glacier which wear out the rugged surface and smoothens them. Plain resulting from the action of these agents of denudation are called peneplains (almost plain) while plains formed from wind action are called pediplains.

Structural Plains

Structural plains are relatively undisturbed horizontal surfaces of the Earth. They are structurally depressed areas of the world that make up some of the most extensive natural lowlands on the Earth's surface.

Plateaus

A plateau also called a high plain or a tableland, is an area of a highland, usually consisting of relatively flat terrain, that is raised significantly above the surrounding area, often with one or more sides with steep slopes. Plateaus can be formed by a number of processes, including upwelling of volcanic magma, extrusion of lava, and erosion by water and glaciers. Plateaus are classified according to their surrounding environment as intermontane, piedmont, or continental.

Formation

Plateaus can be formed by a number of processes, including upwelling of volcanic magma, extrusion of lava, and erosion by water and glaciers.

Volcanic

Volcanic plateaus are produced by volcanic activity. The Columbia Plateau in the northwestern United States is an example. They may be formed by upwelling of volcanic magma or extrusion of lava.

The Pajarito Plateau in New Mexico is an example of a volcanic plateau.

The underlining mechanism in forming plateaus from upwelling starts when magma rises from the mantle, causing the ground to swell upward. In this way, large, flat areas of rock are uplifted to form a plateau. For plateaus formed by extrusion, the rock is built up from lava spreading outward from cracks and weak areas in the crust.

Erosion

Plateaus can also be formed by the erosional processes of glaciers on mountain ranges, leaving them sitting between the mountain ranges. Water can also erode mountains and other landforms down into plateaus. Dissected plateaus are highly eroded plateaus cut by rivers and broken by deep narrow valleys. Computer modeling studies suggest that high plateaus may also be partially a result from the feedback between tectonic deformation and dry climatic conditions created at the lee side of growing orogens.

Classification

Plateaus are classified according to their surrounding environment.

- Intermontane plateaus are the highest in the world, bordered by mountains. The Tibetan Plateau is one such plateau.

- Lava or volcanic plateaus are the plateau that occur in areas of widespread volcanic eruptions. The magma that comes out through narrow cracks or fissures in the crust spread over large area and solidifies. These layers of lava sheets form lava or volcanic plateaus. The Antrim plateau in Northern Ireland, the Deccan Plateau in India and the Columbia Plateau in the United States are examples of lava plateaus.

- Piedmont plateaus are bordered on one side by mountains and on the other by a plain or a sea. The Piedmont Plateau of the Eastern United States between the Appalachian Mountains and the Atlantic Coastal Plain is an example.

- Continental plateaus are bordered on all sides by plains or oceans, forming away from the mountains. An example of a continental plateau is the Antarctic Plateau in East Antarctica.

Glaciers

A glacier is a persistent body of dense ice that is constantly moving under its own weight; it forms where the accumulation of snow exceeds its ablation (melting and sublimation) over many years, often centuries. Glaciers slowly deform and flow due to stresses induced by their weight, creating crevasses, seracs, and other distinguishing features. They also abrade rock and debris from their substrate to create landforms such as cirques and moraines. Glaciers form only on land and are distinct from the much thinner sea ice and lake ice that form on the surface of bodies of water.

On Earth, 99% of glacial ice is contained within vast ice sheets (also known as "continental glaciers") in the polar regions, but glaciers may be found in mountain ranges on every continent including Oceania's high-latitude oceanic island countries such as New Zealand. Between 35°N and 35°S, glaciers occur only in the Himalayas, Andes, Rocky Mountains, a few high mountains in East Africa, Mexico, New Guinea and on Zard Kuh in Iran. Glaciers cover about 10 percent of Earth's land surface. Continental glaciers cover nearly 13 million km² (5 million sq mi) or about 98 percent of Antarctica's 13.2 million km² (5.1 million sq mi), with an average thickness of 2,100 m (7,000 ft). Greenland and Patagonia also have huge expanses of continental glaciers. The volume of glaciers, not including the ice sheets of Antarctica and Greenland, has been estimated as 170,000 km³.

Glacial ice is the largest reservoir of fresh water on Earth. Many glaciers from temperate, alpine and seasonal polar climates store water as ice during the colder seasons and release it later in the form of meltwater as warmer summer temperatures cause the glacier to melt, creating a water source that is especially important for plants, animals and human uses when other sources may be scant. Within high-altitude and Antarctic environments, the seasonal temperature difference is often not sufficient to release meltwater.

Since glacial mass is affected by long-term climatic changes, e.g., precipitation, mean temperature, and cloud cover, glacial mass changes are considered among the most sensitive indicators of climate change and are a major source of variations in sea level.

A large piece of compressed ice, or a glacier, appears blue, as large quantities of water appear blue. This is because water molecules absorb other colors more efficiently than blue. The other reason for the blue color of glaciers is the lack of air bubbles. Air bubbles, which give a white color to ice, are squeezed out by pressure increasing the density of the created ice.

Types

Classification by Size, Shape and Behavior

Mouth of the Schlatenkees Glacier near Innergschlöß, Austria.

Glaciers are categorized by their morphology, thermal characteristics, and behavior. Alpine glaciers form on the crests and slopes of mountains. A glacier that fills a valley is called a valley glacier, or alternatively an alpine glacier or mountain glacier. A large body of glacial ice astride a mountain, mountain range, or volcano is termed an ice cap or ice field. Ice caps have an area less than 50,000 km2 (19,000 sq mi) by definition.

Glacial bodies larger than 50,000 km2 (19,000 sq mi) are called ice sheets or continental glaciers. Several kilometers deep, they obscure the underlying topography. Only nunataks protrude from their surfaces. The only extant ice sheets are the two that cover most of Antarctica and Greenland. They contain vast quantities of fresh water, enough that if both melted, global sea levels would rise by over 70 m (230 ft). Portions of an ice sheet or cap that extend into water are called ice shelves; they tend to be thin with limited slopes and reduced velocities. Narrow, fast-moving sections of an ice sheet are called ice streams. In Antarctica, many ice streams drain into large ice shelves. Some drain directly into the sea, often with an ice tongue, like Mertz Glacier.

The *Grotta del Gelo* is a cave of Etna volcano, the southernmost glacier in Europe.

Sightseeing boat in front of a tidewater glacier, Kenai Fjords National Park, Alaska.

Tidewater glaciers are glaciers that terminate in the sea, including most glaciers flowing from Greenland, Antarctica, Baffin and Ellesmere Islands in Canada, Southeast Alaska, and the Northern and Southern Patagonian Ice Fields. As the ice reaches the sea, pieces break off, or calve, forming icebergs. Most tidewater glaciers calve above sea level, which often results in a tremendous impact as the iceberg strikes the water. Tidewater glaciers undergo centuries-long cycles of advance and retreat that are much less affected by the climate change than those of other glaciers.

Classification by Thermal State

Thermally, a temperate glacier is at melting point throughout the year, from its surface to its base. The ice of a polar glacier is always below the freezing point from the surface to its base, although the surface snowpack may experience seasonal melting. A subpolar glacier includes both temperate and polar ice, depending on depth beneath the surface and position along the length of the glacier. In a similar way, the thermal regime of a glacier is often described by its basal temperature. A cold-based glacier is below freezing at the ice-ground interface, and is thus frozen to the underlying substrate. A warm-based glacier is above or at freezing at the interface, and is able to slide at this contact. This contrast is thought to a large extent to govern the ability of a glacier to effectively erode its bed, as sliding ice promotes plucking at rock from the surface below. Glaciers which are partly cold-based and partly warm-based are known as polythermal.

Formation

Gorner Glacier in Switzerland.

Glaciers form where the accumulation of snow and ice exceeds ablation. A glacier usually originates from a landform called 'cirque' (or corrie or cwm) – a typically armchair-shaped geological feature (such as a depression between mountains enclosed by arêtes) – which collects and compresses through gravity the snow that falls into it. This snow collects and is compacted by the weight of the snow falling above it, forming névé. Further crushing of the individual snowflakes and squeezing the air from the snow turns it into "glacial ice". This glacial ice will fill the cirque until it "overflows" through a geological weakness or vacancy, such as the gap between two mountains. When the mass of snow and ice is sufficiently thick, it begins to move due to a combination of surface slope, gravity and pressure. On steeper slopes, this can occur with as little as 15 m (50 ft) of snow-ice.

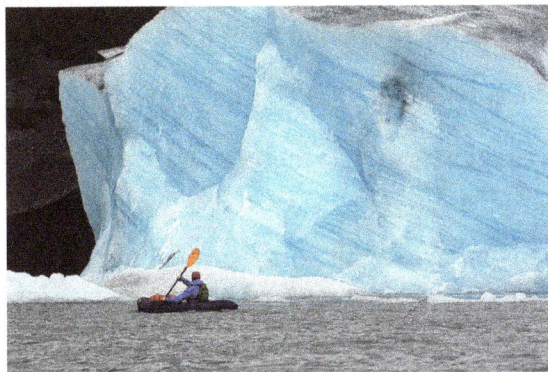

A packrafter passes a wall of freshly exposed blue ice on Spencer Glacier, in Alaska. Glacial ice acts like a filter on light, and the more time light can spend traveling through ice, the bluer it becomes.

In temperate glaciers, snow repeatedly freezes and thaws, changing into granular ice called firn. Under the pressure of the layers of ice and snow above it, this granular ice fuses into denser and denser firn. Over a period of years, layers of firn undergo further compaction and become glacial ice. Glacier ice is slightly less dense than ice formed from frozen water because it contains tiny trapped air bubbles.

Glacial ice has a distinctive blue tint because it absorbs some red light due to an overtone of the infrared OH stretching mode of the water molecule. Liquid water is blue for the same reason. The blue of glacier ice is sometimes misattributed to Rayleigh scattering due to bubbles in the ice.

A glacier cave located on the Perito Moreno Glacier in Argentina.

Structure

A glacier originates at a location called its glacier head and terminates at its glacier foot, snout, or terminus.

Glaciers are broken into zones based on surface snowpack and melt conditions. The ablation zone is the region where there is a net loss in glacier mass. The equilibrium line separates the ablation zone and the accumulation zone; it is the altitude where the amount of new snow gained by accumulation is equal to the amount of ice lost through ablation. The upper part of a glacier, where accumulation exceeds ablation, is called the accumulation zone. In general, the accumulation zone accounts for 60–70% of the glacier's surface area, more if the glacier calves icebergs. Ice in the accumulation zone is deep enough to exert a downward force that erodes underlying rock. After a glacier melts, it often leaves behind a bowl- or amphitheater-shaped depression that ranges in size from large basins like the Great Lakes to smaller mountain depressions known as cirques.

The accumulation zone can be subdivided based on its melt conditions:

- The dry snow zone is a region where no melt occurs, even in the summer, and the snowpack remains dry.

- The percolation zone is an area with some surface melt, causing meltwater to percolate into the snowpack. This zone is often marked by refrozen ice lenses, glands, and layers. The snowpack also never reaches melting point.

- Near the equilibrium line on some glaciers, a superimposed ice zone develops. This zone is where meltwater refreezes as a cold layer in the glacier, forming a continuous mass of ice.

- The wet snow zone is the region where all of the snow deposited since the end of the previous summer has been raised to 0 °C.

The health of a glacier is usually assessed by determining the glacier mass balance or observing terminus behavior. Healthy glaciers have large accumulation zones, more than 60% of their area snowcovered at the end of the melt season, and a terminus with vigorous flow.

Following the Little Ice Age's end around 1850, glaciers around the Earth have retreated substantially. A slight cooling led to the advance of many alpine glaciers between 1950 and 1985, but since 1985 glacier retreat and mass loss has become larger and increasingly ubiquitous.

Motion

Shear or herring-bone crevasses on Emmons Glacier (Mount Rainier); such crevasses often form near the edge of a glacier where interactions with underlying or marginal rock impede flow. In this case, the impediment appears to be some distance from the near margin of the glacier.

Glaciers move, or flow, downhill due to gravity and the internal deformation of ice. Ice behaves like a brittle solid until its thickness exceeds about 50 m (160 ft). The pressure on ice deeper than 50 m causes plastic flow. At the molecular level, ice consists of stacked layers of molecules with relatively weak bonds between layers. When the stress on the layer above exceeds the inter-layer binding strength, it moves faster than the layer below.

Glaciers also move through basal sliding. In this process, a glacier slides over the terrain on which it sits, lubricated by the presence of liquid water. The water is created from ice that melts under high pressure from frictional heating. Basal sliding is dominant in temperate, or warm-based glaciers.

Although evidence in favour of glacial flow was known by the early 19th century, other theories of glacial motion were advanced, such as the idea that melt water, refreezing inside glaciers, caused the glacier to dilate and extend its length. As it became clear that glaciers behaved to some degree as if the ice were a viscous fluid, it was argued that "regelation", or the melting and refreezing of ice at a temperature lowered by the pressure on the ice inside the glacier, was what allowed the ice to deform and flow. James Forbes came up with the essentially correct explanation in the 1840s, although it was several decades before it was fully accepted.

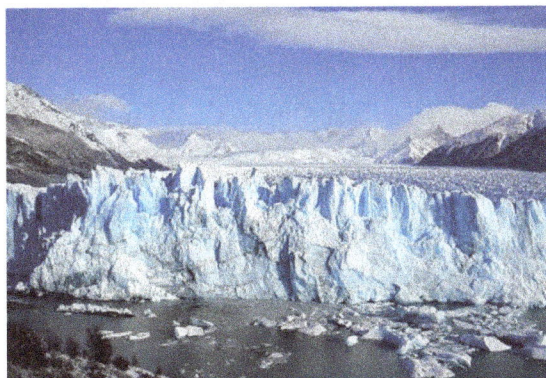

Perito Moreno glacier.

Fracture Zone and Cracks

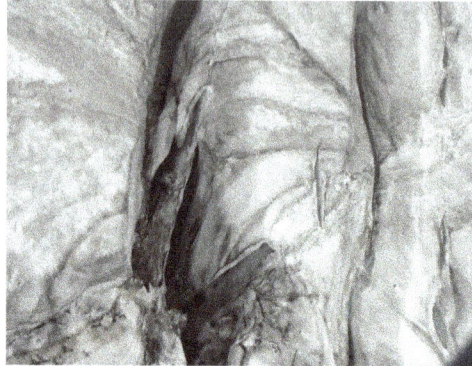

Ice cracks in the Titlis Glacier.

The top 50 m (160 ft) of a glacier are rigid because they are under low pressure. This upper section is known as the fracture zone and moves mostly as a single unit over the plastically flowing lower section. When a glacier moves through irregular terrain, cracks called crevasses develop in the fracture zone. Crevasses form due to differences in glacier velocity. If two rigid sections of a glacier move at different speeds and directions, shear forces cause them to break apart, opening a crevasse. Crevasses are seldom more than 46 m (150 ft) deep but in some cases can be 300 m (1,000 ft) or even deeper. Beneath this point, the plasticity of the ice is too great for cracks to form. Intersecting crevasses can create isolated peaks in the ice, called seracs.

Crevasses can form in several different ways. Transverse crevasses are transverse to flow and form where steeper slopes cause a glacier to accelerate. Longitudinal crevasses form semi-parallel to flow where a glacier expands laterally. Marginal crevasses form from the edge of the glacier, due to the reduction in speed caused by friction of the valley walls. Marginal crevasses are usually largely transverse to flow. Moving glacier ice can sometimes separate from stagnant ice above, forming a bergschrund. Bergschrunds resemble crevasses but are singular features at a glacier's margins.

Crevasses make travel over glaciers hazardous, especially when they are hidden by fragile snow bridges.

Crossing a crevasse on the Easton Glacier, Mount Baker,
in the North Cascades, United States.

Below the equilibrium line, glacial meltwater is concentrated in stream channels. Meltwater can pool in proglacial lakes on top of a glacier or descend into the depths of a glacier via moulins. Streams within or beneath a glacier flow in englacial or sub-glacial tunnels. These tunnels sometimes reemerge at the glacier's surface.

Speed

The speed of glacial displacement is partly determined by friction. Friction makes the ice at the bottom of the glacier move more slowly than ice at the top. In alpine glaciers, friction is also generated at the valley's side walls, which slows the edges relative to the center.

Mean speeds vary greatly, but is typically around 1 m (3 ft) per day. There may be no motion in stagnant areas; for example, in parts of Alaska, trees can establish themselves on surface sediment deposits. In other cases, glaciers can move as fast as 20–30 m (70–100 ft) per day, such as in Greenland's Jakobshavn Isbræ. Velocity increases with increasing slope, increasing thickness, increasing snowfall, increasing longitudinal confinement, increasing basal temperature, increasing meltwater production and reduced bed hardness.

A few glaciers have periods of very rapid advancement called surges. These glaciers exhibit normal movement until suddenly they accelerate, then return to their previous state. During these surges, the glacier may reach velocities far greater than normal speed. These surges may be caused by failure of the underlying bedrock, the pooling of meltwater at the base of the glacier — perhaps delivered from a supraglacial lake — or the simple accumulation of mass beyond a critical "tipping point". Temporary rates up to 90 m (300 ft) per day have occurred when increased temperature or overlying pressure caused bottom ice to melt and water to accumulate beneath a glacier.

In glaciated areas where the glacier moves faster than one km per year, glacial earthquakes occur. These are large scale earthquakes that have seismic magnitudes as high as 6.1. The number of glacial earthquakes in Greenland peaks every year in July, August and September and increased rapidly in the 1990s and 2000s. In a study using data from January 1993 through October 2005, more events were detected every year since 2002, and twice as many events were recorded in 2005 as there were in any other year.

Ogives

Ogives (or Forbes bands) are alternating wave crests and valleys that appear as dark and light bands of ice on glacier surfaces. They are linked to seasonal motion of glaciers; the width of one dark and one light band generally equals the annual movement of the glacier. Ogives are formed when ice from an icefall is severely broken up, increasing ablation surface area during summer. This creates a swale and space for snow accumulation in the winter, which in turn creates a ridge. Sometimes ogives consist only of undulations or color bands and are described as wave ogives or band ogives.

Geography

Glaciers are present on every continent and approximately fifty countries, excluding those (Australia, South Africa) that have glaciers only on distant subantarctic island territories. Extensive glaciers are found in Antarctica, Argentina, Chile, Canada, Alaska, Greenland and Iceland. Mountain glaciers are widespread, especially in the Andes, the Himalayas, the Rocky Mountains, the

Caucasus, Scandinavian mountains, and the Alps. Snezhnika glacier in Pirin Mountain, Bulgaria with a latitude of 41°46′09″ N is the southernmost glacial mass in Europe. Mainland Australia currently contains no glaciers, although a small glacier on Mount Kosciuszko was present in the last glacial period. In New Guinea, small, rapidly diminishing, glaciers are located on its highest summit massif of Puncak Jaya. Africa has glaciers on Mount Kilimanjaro in Tanzania, on Mount Kenya and in the Rwenzori Mountains. Oceanic islands with glaciers include Iceland, several of the islands off the coast of Norway including Svalbard and Jan Mayen to the far North, New Zealand and the subantarctic islands of Marion, Heard, Grande Terre (Kerguelen) and Bouvet. During glacial periods of the Quaternary, Taiwan, Hawaii on Mauna Kea and Tenerife also had large alpine glaciers, while the Faroe and Crozet Islands were completely glaciated.

Black ice glacier near Aconcagua, Argentina.

The permanent snow cover necessary for glacier formation is affected by factors such as the degree of slope on the land, amount of snowfall and the winds. Glaciers can be found in all latitudes except from 20° to 27° north and south of the equator where the presence of the descending limb of the Hadley circulation lowers precipitation so much that with high insolation snow lines reach above 6,500 m (21,330 ft). Between 19°N and 19°S, however, precipitation is higher and the mountains above 5,000 m (16,400 ft) usually have permanent snow.

Even at high latitudes, glacier formation is not inevitable. Areas of the Arctic, such as Banks Island, and the McMurdo Dry Valleys in Antarctica are considered polar deserts where glaciers cannot form because they receive little snowfall despite the bitter cold. Cold air, unlike warm air, is unable to transport much water vapor. Even during glacial periods of the Quaternary, Manchuria, lowland Siberia, and central and northern Alaska, though extraordinarily cold, had such light snowfall that glaciers could not form.

In addition to the dry, unglaciated polar regions, some mountains and volcanoes in Bolivia, Chile and Argentina are high (4,500 to 6,900 m or 14,800 to 22,600 ft) and cold, but the relative lack of precipitation prevents snow from accumulating into glaciers. This is because these peaks are located near or in the hyperarid Atacama Desert.

Glacial Geology

As glaciers flow over bedrock, they soften and lift blocks of rock into the ice. This process, called plucking, is caused by subglacial water that penetrates fractures in the bedrock and subsequently freezes and expands. This expansion causes the ice to act as a lever that loosens the rock by lifting it. Thus, sediments of all sizes become part of the glacier's load. If a retreating glacier gains enough debris, it may become a rock glacier, like the Timpanogos Glacier in Utah.

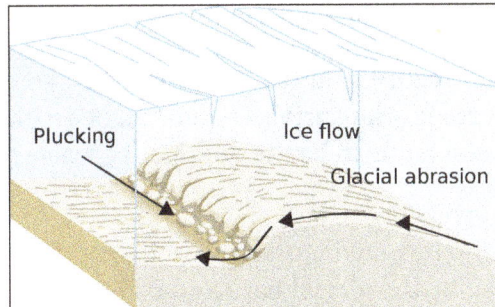

Diagram of glacial plucking and abrasion.

Glaciers erode terrain through two principal processes: abrasion and plucking.

Abrasion occurs when the ice and its load of rock fragments slide over bedrock and function as sandpaper, smoothing and polishing the bedrock below. The pulverized rock this process produces is called rock flour and is made up of rock grains between 0.002 and 0.00625 mm in size. Abrasion leads to steeper valley walls and mountain slopes in alpine settings, which can cause avalanches and rock slides, which add even more material to the glacier.

Glacial abrasion is commonly characterized by glacial striations. Glaciers produce these when they contain large boulders that carve long scratches in the bedrock. By mapping the direction of the striations, researchers can determine the direction of the glacier's movement. Similar to striations are chatter marks, lines of crescent-shape depressions in the rock underlying a glacier. They are formed by abrasion when boulders in the glacier are repeatedly caught and released as they are dragged along the bedrock.

Glacially plucked granitic bedrock near Mariehamn, Åland Islands.

The rate of glacier erosion varies. Six factors control erosion rate:

- Velocity of glacial movement,

- Thickness of the ice,

- Shape, abundance and hardness of rock fragments contained in the ice at the bottom of the glacier,

- Relative ease of erosion of the surface under the glacier,

- Thermal conditions at the glacier base,

- Permeability and water pressure at the glacier base.

When the bedrock has frequent fractures on the surface, glacial erosion rates tend to increase as plucking is the main erosive force on the surface; when the bedrock has wide gaps between sporadic fractures, however, abrasion tends to be the dominant erosive form and glacial erosion rates become slow.

Glaciers in lower latitudes tend to be much more erosive than glaciers in higher latitudes, because they have more meltwater reaching the glacial base and facilitate sediment production and transport under the same moving speed and amount of ice.

Material that becomes incorporated in a glacier is typically carried as far as the zone of ablation before being deposited. Glacial deposits are of two distinct types:

- Glacial till: material directly deposited from glacial ice. Till includes a mixture of undifferentiated material ranging from clay size to boulders, the usual composition of a moraine.

- Fluvial and outwash sediments: sediments deposited by water. These deposits are stratified by size.

Larger pieces of rock that are encrusted in till or deposited on the surface are called "glacial erratics". They range in size from pebbles to boulders, but as they are often moved great distances, they may be drastically different from the material upon which they are found. Patterns of glacial erratics hint at past glacial motions.

Moraines

Glacial moraines above Lake Louise, Alberta, Canada.

Glacial moraines are formed by the deposition of material from a glacier and are exposed after the glacier has retreated. They usually appear as linear mounds of till, a non-sorted mixture of rock, gravel and boulders within a matrix of a fine powdery material. Terminal or end moraines are formed at the foot or terminal end of a glacier. Lateral moraines are formed on the sides of the glacier. Medial moraines are formed when two different glaciers merge and the lateral moraines of each coalesce to form a moraine in the middle of the combined glacier. Less apparent are ground moraines, also called glacial drift, which often blankets the surface underneath the glacier downslope from the equilibrium line.

The term moraine is of French origin. It was coined by peasants to describe alluvial embankments and rims found near the margins of glaciers in the French Alps. In modern geology, the term is used more broadly, and is applied to a series of formations, all of which are composed of till. Moraines can also create moraine dammed lakes.

Drumlins

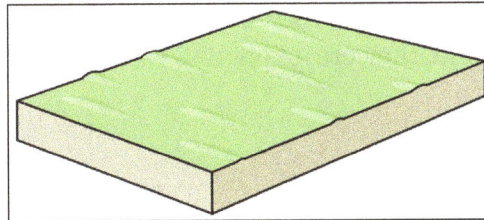

A drumlin field forms after a glacier has modified the landscape. The teardrop-shaped formations denote the direction of the ice flow.

Drumlins are asymmetrical, canoe shaped hills made mainly of till. Their heights vary from 15 to 50 meters and they can reach a kilometer in length. The steepest side of the hill faces the direction from which the ice advanced (*stoss*), while a longer slope is left in the ice's direction of movement (*lee*).

Drumlins are found in groups called drumlin fields or drumlin camps. One of these fields is found east of Rochester, New York; it is estimated to contain about 10,000 drumlins.

Although the process that forms drumlins is not fully understood, their shape implies that they are products of the plastic deformation zone of ancient glaciers. It is believed that many drumlins were formed when glaciers advanced over and altered the deposits of earlier glaciers.

Glacial Valleys, Cirques, Arêtes and Pyramidal Peaks

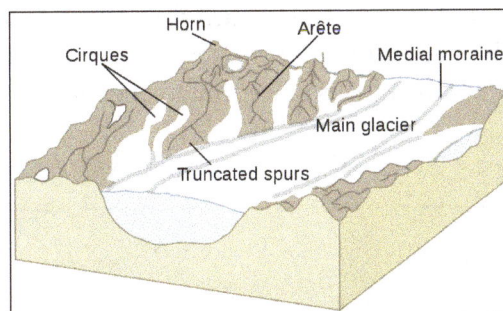

Features of a glacial landscape.

Before glaciation, mountain valleys have a characteristic "V" shape, produced by eroding water. During glaciation, these valleys are often widened, deepened and smoothed to form a "U"-shaped glacial valley or glacial trough, as it is sometimes called. The erosion that creates glacial valleys truncates any spurs of rock or earth that may have earlier extended across the valley, creating broadly triangular-shaped cliffs called truncated spurs. Within glacial valleys, depressions created by plucking and abrasion can be filled by lakes, called paternoster lakes. If a glacial valley runs into a large body of water, it forms a fjord.

Typically glaciers deepen their valleys more than their smaller tributaries. Therefore, when glaciers recede, the valleys of the tributary glaciers remain above the main glacier's depression and are called hanging valleys.

At the start of a classic valley glacier is a bowl-shaped cirque, which has escarped walls on three sides but is open on the side that descends into the valley. Cirques are where ice begins to accumulate in a glacier. Two glacial cirques may form back to back and erode their backwalls until only a narrow ridge, called an arête is left. This structure may result in a mountain pass. If multiple cirques encircle a single mountain, they create pointed pyramidal peaks; particularly steep examples are called horns.

Roches Moutonnées

Passage of glacial ice over an area of bedrock may cause the rock to be sculpted into a knoll called a *roche moutonnée,* or "sheepback" rock. Roches moutonnées may be elongated, rounded and asymmetrical in shape. They range in length from less than a meter to several hundred meters long. Roches moutonnées have a gentle slope on their up-glacier sides and a steep to vertical face on their down-glacier sides. The glacier abrades the smooth slope on the upstream side as it flows along, but tears rock fragments loose and carries them away from the downstream side via plucking.

Alluvial Stratification

As the water that rises from the ablation zone moves away from the glacier, it carries fine eroded sediments with it. As the speed of the water decreases, so does its capacity to carry objects in suspension. The water thus gradually deposits the sediment as it runs, creating an alluvial plain. When this phenomenon occurs in a valley, it is called a *valley train*. When the deposition is in an estuary, the sediments are known as bay mud.

Outwash plains and valley trains are usually accompanied by basins known as "kettles". These are small lakes formed when large ice blocks that are trapped in alluvium melt and produce water-filled depressions. Kettle diameters range from 5 m to 13 km, with depths of up to 45 meters. Most are circular in shape because the blocks of ice that formed them were rounded as they melted.

Glacial Deposits

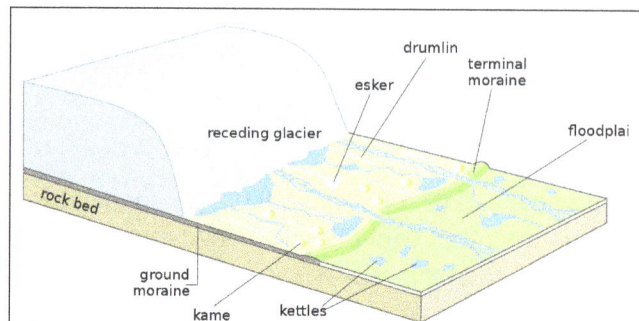

Landscape produced by a receding glacier.

When a glacier's size shrinks below a critical point, its flow stops and it becomes stationary. Meanwhile, meltwater within and beneath the ice leaves stratified alluvial deposits. These deposits, in the forms of columns, terraces and clusters, remain after the glacier melts and are known as "glacial deposits".

Glacial deposits that take the shape of hills or mounds are called *kames*. Some kames form when meltwater deposits sediments through openings in the interior of the ice. Others are produced by fans or deltas created by meltwater. When the glacial ice occupies a valley, it can form terraces or kames along the sides of the valley.

Long, sinuous glacial deposits are called *eskers*. Eskers are composed of sand and gravel that was deposited by meltwater streams that flowed through ice tunnels within or beneath a glacier. They remain after the ice melts, with heights exceeding 100 meters and lengths of as long as 100 km.

Loess Deposits

Very fine glacial sediments or rock flour is often picked up by wind blowing over the bare surface and may be deposited great distances from the original fluvial deposition site. These eolian loess deposits may be very deep, even hundreds of meters, as in areas of China and the Midwestern United States of America. Katabatic winds can be important in this process.

Isostatic Rebound

Large masses, such as ice sheets or glaciers, can depress the crust of the Earth into the mantle. The depression usually totals a third of the ice sheet or glacier's thickness. After the ice sheet or glacier melts, the mantle begins to flow back to its original position, pushing the crust back up. This post-glacial rebound, which proceeds very slowly after the melting of the ice sheet or glacier, is currently occurring in measurable amounts in Scandinavia and the Great Lakes region of North America.

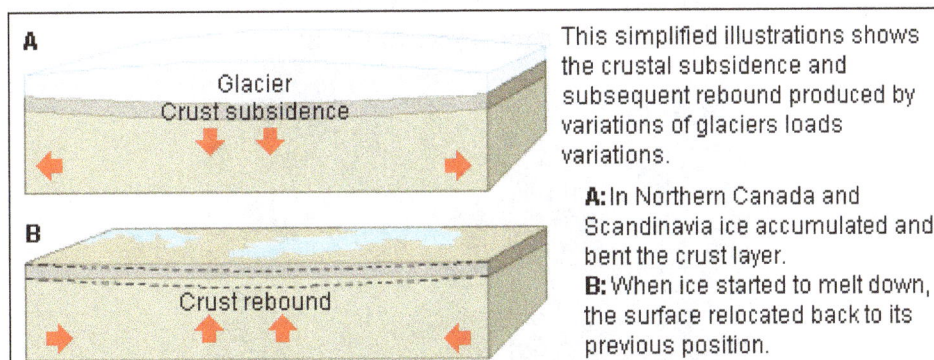

This simplified illustrations shows the crustal subsidence and subsequent rebound produced by variations of glaciers loads variations.

A: In Northern Canada and Scandinavia ice accumulated and bent the crust layer.
B: When ice started to melt down, the surface relocated back to its previous position.

Isostatic pressure by a glacier on the Earth's crust.

A geomorphological feature created by the same process on a smaller scale is known as *dilation-faulting*. It occurs where previously compressed rock is allowed to return to its original shape more rapidly than can be maintained without faulting. This leads to an effect similar to what would be seen if the rock were hit by a large hammer. Dilation faulting can be observed in recently de-glaciated parts of Iceland and Cumbria.

Canyon

A canyon or gorge is a deep cleft between escarpments or cliffs resulting from weathering and the erosive activity of a river over geologic timescales. Rivers have a natural tendency to cut through underlying surfaces, eventually wearing away rock layers as sediments are removed downstream. A river bed will gradually reach a baseline elevation, which is the same elevation as the body of water into which the river drains. The processes of weathering and erosion will form canyons when the river's headwaters and estuary are at significantly different elevations, particularly through regions where softer rock layers are intermingled with harder layers more resistant to weathering.

A canyon may also refer to a rift between two mountain peaks, such as those in ranges including the Rocky Mountains, the Alps, the Himalayas or the Andes. Usually a river or stream carve out such splits between mountains. Examples of mountain-type canyons are Provo Canyon in Utah or Yosemite Valley in California's Sierra Nevada. Canyons within mountains, or gorges that have an opening on only one side, are called box canyons. Slot canyons are very narrow canyons that often have smooth walls.

Steep-sided valleys in the seabed of the continental slope are referred to as submarine canyons. Unlike canyons on land, submarine canyons are thought to be formed by turbidity currents and landslides.

Formation

Most canyons were formed by a process of long-time erosion from a plateau or table-land level. The cliffs form because harder rock strata that are resistant to erosion and weathering remain exposed on the valley walls.

Canyons are much more common in arid than in wet areas because physical weathering has a more localized effect in arid zones. The wind and water from the river combine to erode and cut away less resistant materials such as shales. The freezing and expansion of water also serves to help form canyons. Water seeps into cracks between the rocks and freezes, pushing the rocks apart and eventually causing large chunks to break off the canyon walls, in a process known as frost wedging. Canyon walls are often formed of resistant sandstones or granite.

Snake River Canyon, Idaho.

Sometimes large rivers run through canyons as the result of gradual geological uplift. These are called entrenched rivers, because they are unable to easily alter their course. In the United States, the Colorado River in the Southwest and the Snake River in the Northwest are two examples of tectonic uplift.

Canyons often form in areas of limestone rock. As limestone is soluble to a certain extent, cave systems form in the rock. When these collapse, a canyon is left, as in the Mendip Hills in Somerset and Yorkshire Dales in Yorkshire, England.

Box Canyon

A box canyon is a small canyon that is generally shorter and narrower than a river canyon, with steep walls on three sides, allowing access and egress only through the mouth of the canyon. Box canyons were frequently used in the western United States as convenient corrals, with their entrances fenced.

Largest Canyons

The definition of "largest canyon" is imprecise, because a canyon can be large by its depth, its length, or the total area of the canyon system. Also, the inaccessibility of the major canyons in the Himalaya contributes to their not being regarded as candidates for the biggest canyon. The definition of "deepest canyon" is similarly imprecise, especially if one includes mountain canyons as well as canyons cut through relatively flat plateaus (which have a somewhat well-defined rim elevation).

The Yarlung Tsangpo Grand Canyon (or Tsangpo Canyon), along the Yarlung Tsangpo River in Tibet, is regarded by some as the deepest canyon in the world at 5,500 m (18,000 ft). It is slightly longer than the Grand Canyon in the United States. Others consider the Kali Gandaki Gorge in midwest Nepal to be the deepest canyon, with a 6400 m (21,000 ft) difference between the level of the river and the peaks surrounding it.

Vying for deepest canyon in the Americas are the Cotahuasi Canyon and Colca Canyon, in southern Peru. Both have been measured at over 3500 m (12,000 ft) deep.

The Grand Canyon of northern Arizona in the United States, with an average depth of 1,600 m (one mile) and a volume of 4.17 trillion cubic metres, is one of the world's largest canyons. It was among the 28 finalists of the New 7 Wonders of Nature worldwide poll. (Some referred to it as one of the 7 natural wonders of the world.)

The largest canyon in Africa is the Fish River Canyon in Namibia.

In August 2013, the discovery of Greenland's Grand Canyon was reported, based on the analysis of data from Operation IceBridge. It is located under an ice sheet. At 750 kilometres (466 mi) long, it is believed to be the longest canyon in the world.

The Capertee Valley in Australia is commonly reported as being the second largest (in terms of width) canyon in the world.

Panoramic view of the Capertee Valley in Australia, the second largest (in terms of width) of any canyon in the world.

Volcanoes

A volcano is a rupture in the crust of a planetary-mass object, such as Earth, that allows hot lava, volcanic ash, and gases to escape from a magma chamber below the surface.

Sabancaya volcano, Peru.

Cordillera de Apaneca volcanic range in El Salvador.

Earth's volcanoes occur because its crust is broken into 17 major, rigid tectonic plates that float on a hotter, softer layer in its mantle. Therefore, on Earth, volcanoes are generally found where tectonic plates are diverging or converging, and most are found underwater. For example, a mid-oceanic ridge, such as the Mid-Atlantic Ridge, has volcanoes caused by divergent tectonic plates whereas the Pacific Ring of Fire has volcanoes caused by convergent tectonic plates. Volcanoes can also form where there is stretching and thinning of the crust's plates, e.g., in the East African Rift and the Wells Gray-Clearwater volcanic field and Rio Grande Rift in North America. This type of volcanism falls under the umbrella of "plate hypothesis" volcanism. Volcanism away from plate boundaries has also been explained as mantle plumes. These so-called "hotspots", for example Hawaii, are postulated to arise from upwelling diapirs with magma from the core–mantle boundary, 3,000 km deep in the Earth. Volcanoes are usually not created where two tectonic plates slide past one another.

The country is home to 170 volcanoes, 23 which are active, including two calderas, one being a supervolcano. El Salvador has earned the epithets endearment La Tierra de Soberbios Volcanes, (The Land of Magnificent Volcanoes).

Cleveland Volcano in the Aleutian Islands of Alaska photographed from the International Space Station.

Erupting volcanoes can pose many hazards, not only in the immediate vicinity of the eruption. One such hazard is that volcanic ash can be a threat to aircraft, in particular those with jet engines where ash particles can be melted by the high operating temperature; the melted particles then adhere to the turbine blades and alter their shape, disrupting the operation of the turbine. Large eruptions can affect temperature as ash and droplets of sulfuric acid obscure the sun and cool the Earth's lower atmosphere (or troposphere); however, they also absorb heat radiated from the Earth, thereby warming the upper atmosphere (or stratosphere). Historically, volcanic winters have caused catastrophic famines.

Fountain of lava erupting from a volcanic cone in Hawaii.

Satellite image of Mount Shasta in California.

Plate Tectonics

Divergent Plate Boundaries

At the mid-oceanic ridges, two tectonic plates diverge from one another as new oceanic crust is formed by the cooling and solidifying of hot molten rock. Because the crust is very thin at these ridges due to the pull of the tectonic plates, the release of pressure leads to adiabatic expansion (without transfer of heat or matter) and the partial melting of the mantle, causing volcanism and creating new oceanic crust. Most divergent plate boundaries are at the bottom of the oceans; therefore, most volcanic activity on the Earth is submarine, forming new seafloor. Black smokers (also known as deep sea vents) are evidence of this kind of volcanic activity. Where the mid-oceanic ridge is above sea-level, volcanic islands are formed; for example, Iceland.

Map showing the divergent plate boundaries (oceanic spreading ridges) and recent sub-aerial volcanoes.

Convergent Plate Boundaries

Subduction zones are places where two plates, usually an oceanic plate and a continental plate, collide. In this case, the oceanic plate subducts, or submerges, under the continental plate, forming a deep ocean trench just offshore. In a process called flux melting, water released from the subducting plate lowers the melting temperature of the overlying mantle wedge, thus creating magma. This magma tends to be extremely viscous because of its high silica content, so it often does not attain the surface but cools and solidifies at depth. When it does reach the surface, however, a volcano is formed. Typical examples are Mount Etna and the volcanoes in the Pacific Ring of Fire.

Hotspots

Hotspots are volcanic areas believed to be formed by mantle plumes, which are hypothesized to be columns of hot material rising from the core-mantle boundary in a fixed space that causes large-volume melting. Because tectonic plates move across them, each volcano becomes dormant and is eventually re-formed as the plate advances over the postulated plume. The Hawaiian Islands are said to have been formed in such a manner; so has the Snake River Plain, with the Yellowstone Caldera being the part of the North American plate above the hot spot. This theory, however, has been doubted.

Volcanic Features

Lakagigar fissure vent in Iceland, the source of the major world climate alteration of 1783–84, has a chain of volcanic cones along its length.

Skjaldbreiður, a shield volcano whose name means "broad shield".

The most common perception of a volcano is of a conical mountain, spewing lava and poisonous gases from a crater at its summit; however, this describes just one of the many types of volcano. The features of volcanoes are much more complicated and their structure and behavior depends on a number of factors. Some volcanoes have rugged peaks formed by lava domes rather than a summit crater while others have landscape features such as massive plateaus. Vents that issue volcanic material (including lava and ash) and gases (mainly steam and magmatic gases) can develop anywhere on the landform and may give rise to smaller cones such as Puʻu ʻŌʻō on a flank of Hawaii's Kīlauea. Other types of volcano include cryovolcanoes (or ice volcanoes), particularly on some moons of Jupiter, Saturn, and Neptune; and mud volcanoes, which are formations often not associated with known magmatic activity. Active mud volcanoes tend to involve temperatures much lower than those of igneous volcanoes except when the mud volcano is actually a vent of an igneous volcano.

Fissure Vents

Volcanic fissure vents are flat, linear fractures through which lava emerges.

Shield Volcanoes

Shield volcanoes, so named for their broad, shield-like profiles, are formed by the eruption of low-viscosity lava that can flow a great distance from a vent. They generally do not explode catastrophically. Since low-viscosity magma is typically low in silica, shield volcanoes are more common in oceanic than continental settings. The Hawaiian volcanic chain is a series of shield cones, and they are common in Iceland, as well.

Lava Domes

Lava domes are built by slow eruptions of highly viscous lava. They are sometimes formed within the crater of a previous volcanic eruption, as in the case of Mount Saint Helens, but can also form independently, as in the case of Lassen Peak. Like stratovolcanoes, they can produce violent, explosive eruptions, but their lava generally does not flow far from the originating vent.

Cryptodomes

Cryptodomes are formed when viscous lava is forced upward causing the surface to bulge. The 1980 eruption of Mount St. Helens was an example; lava beneath the surface of the mountain created an upward bulge which slid down the north side of the mountain.

Volcanic Cones (Cinder Cones)

Izalco volcano, the youngest volcano in El Salvador. Izalco erupted almost continuously from 1770 (when it formed) to 1958, earning it the nickname of "Lighthouse of the Pacific".

Volcanic cones or cinder cones result from eruptions of mostly small pieces of scoria and pyroclastics (both resemble cinders, hence the name of this volcano type) that build up around the vent. These can be relatively short-lived eruptions that produce a cone-shaped hill perhaps 30 to 400 meters high. Most cinder cones erupt only once. Cinder cones may form as flank vents on larger volcanoes, or occur on their own. Parícutin in Mexico and Sunset Crater in Arizona are examples of cinder cones. In New Mexico, Caja del Rio is a volcanic field of over 60 cinder cones.

Based on satellite images it was suggested that cinder cones might occur on other terrestrial bodies in the Solar system too; on the surface of Mars and the Moon.

Stratovolcanoes (Composite Volcanoes)

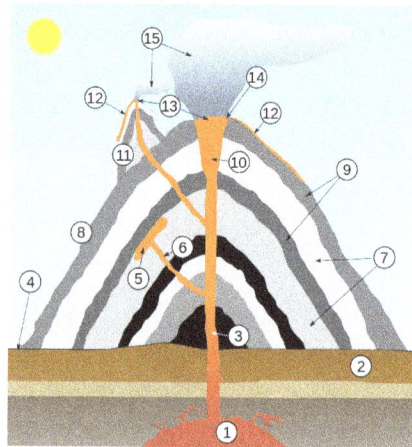

Cross-section through a stratovolcano (vertical scale is exaggerated):

1. Large magma chamber,

2. Bedrock,

3. Conduit (pipe),

4. Base,

5. Sill,

6. Dike,

7. Layers of ash emitted by the volcano,

8. Flank,

9. Layers of lava emitted by the volcano,

10. Throat,

11. Parasitic cone,

12. Lava flow,

13. Vent,

14. Crater,

15. Ash cloud.

Stratovolcanoes or composite volcanoes are tall conical mountains composed of lava flows and other ejecta in alternate layers, the strata that gives rise to the name. Stratovolcanoes are also known as composite volcanoes because they are created from multiple structures during different kinds of eruptions. Strato/composite volcanoes are made of cinders, ash, and lava. Cinders and ash pile on top of each other, lava flows on top of the ash, where it cools and hardens, and then the process repeats. Classic examples include Mount Fuji in Japan, Mayon Volcano in the Philippines, and Mount Vesuvius and Stromboli in Italy.

Throughout recorded history, ash produced by the explosive eruption of stratovolcanoes has posed the greatest volcanic hazard to civilizations. Not only do stratovolcanoes have greater pressure build-up from the underlying lava flow than shield volcanoes, but their fissure vents and monogenetic volcanic fields (volcanic cones) also have more powerful eruptions because they are often under extension. They are also steeper than shield volcanoes, with slopes of 30–35° compared to slopes of generally 5–10°, and their loose tephra are material for dangerous lahars. Large pieces of tephra are called volcanic bombs. Big bombs can measure more than 4 feet (1.2 meters) across and weigh several tons.

Supervolcanoes

A supervolcano usually has a large caldera and can produce devastation on an enormous, sometimes continental, scale. Such volcanoes are able to severely cool global temperatures for many years after the eruption due to the huge volumes of sulfur and ash released into the atmosphere. They are the most dangerous type of volcano. Examples include Yellowstone Caldera in Yellowstone National Park and Valles Caldera in New Mexico (both western United States); Lake Taupo in New Zealand; Lake Toba in Sumatra, Indonesia; and Ngorongoro Crater in Tanzania. Because of the enormous area they may cover, supervolcanoes are hard to identify centuries after an eruption. Similarly, large igneous provinces are also considered supervolcanoes because of the vast amount of basalt lava erupted (even though the lava flow is non-explosive).

Underwater Volcanoes

Submarine volcanoes are common features of the ocean floor. In shallow water, active volcanoes disclose their presence by blasting steam and rocky debris high above the ocean's surface. In the ocean's deep, the tremendous weight of the water above prevents the explosive release of steam and gases; however, they can be detected by hydrophones and discoloration of water because of volcanic gases. Pillow lava is a common eruptive product of submarine volcanoes and is characterized by thick sequences of discontinuous pillow-shaped masses which form under water. Even large submarine eruptions may not disturb the ocean surface due to the rapid cooling effect and increased buoyancy of water (as compared to air) which often causes volcanic vents to form steep pillars on the ocean floor. Hydrothermal vents are common near these volcanoes, and some support peculiar ecosystems based on dissolved minerals. Over time, the formations created by submarine volcanoes may become so large that they break the ocean surface as new islands or floating pumice rafts.

Subglacial Volcanoes

Subglacial volcanoes develop underneath icecaps. They are made up of flat lava which flows at the top of extensive pillow lavas and palagonite. When the icecap melts, the lava on top collapses, leaving a flat-topped mountain. These volcanoes are also called table mountains, tuyas, or (uncommonly) mobergs. Very good examples of this type of volcano can be seen in Iceland, however, there are also tuyas in British Columbia. The origin of the term comes from Tuya Butte, which is one of the several tuyas in the area of the Tuya River and Tuya Range in northern British Columbia. Tuya Butte was the first such landform analyzed and so its name has entered the geological literature for this kind of volcanic formation. The Tuya Mountains Provincial Park was recently established to protect this unusual landscape, which lies north of Tuya Lake and south of the Jennings River near the boundary with the Yukon Territory.

Mud Volcanoes

Mud volcanoes or mud domes are formations created by geo-excreted liquids and gases, although there are several processes which may cause such activity. The largest structures are 10 kilometers in diameter and reach 700 meters high.

Erupted Material

Pāhoehoe lava flow on Hawaii. The picture shows overflows of a main lava channel.

San Miguel (volcano), El Salvador.

On December 29, 2013, San Miguel volcano, also known as "Chaparrastique", erupted at 10:30 local time, spewing a large column of ash and smoke into the sky; the eruption, the first in 11 years, was seen from space and prompted the evacuation of thousands of people living in a 3 km radius around the volcano.

Lava Composition

Sarychev Peak eruption, Matua Island, oblique satellite view.

Another way of classifying volcanoes is by the *composition of material erupted* (lava), since this affects the shape of the volcano. Lava can be broadly classified into four different compositions:

- If the erupted magma contains a high percentage (>63%) of silica, the lava is called felsic:

  ◦ Felsic lavas (dacites or rhyolites) tend to be highly viscous (not very fluid) and are erupted as domes or short, stubby flows. Viscous lavas tend to form stratovolcanoes or lava domes. Lassen Peak in California is an example of a volcano formed from felsic lava and is actually a large lava dome.

- ○ Because siliceous magmas are so viscous, they tend to trap volatiles (gases) that are present, which cause the magma to erupt catastrophically, eventually forming stratovolcanoes. Pyroclastic flows (ignimbrites) are highly hazardous products of such volcanoes, since they are composed of molten volcanic ash too heavy to go up into the atmosphere, so they hug the volcano's slopes and travel far from their vents during large eruptions. Temperatures as high as 1,200 °C are known to occur in pyroclastic flows, which will incinerate everything flammable in their path and thick layers of hot pyroclastic flow deposits can be laid down, often up to many meters thick. Alaska's Valley of Ten Thousand Smokes, formed by the eruption of Novarupta near Katmai in 1912, is an example of a thick pyroclastic flow or ignimbrite deposit. Volcanic ash that is light enough to be erupted high into the Earth's atmosphere may travel many kilometres before it falls back to ground as a tuff.

- If the erupted magma contains 52–63% silica, the lava is of *intermediate* composition:

  - ○ These "andesitic" volcanoes generally only occur above subduction zones (e.g. Mount Merapi in Indonesia).

  - ○ Andesitic lava is typically formed at convergent boundary margins of tectonic plates, by several processes:

    - ▪ Hydration melting of peridotite and fractional crystallization,

    - ▪ Melting of subducted slab containing sediments,

    - ▪ Magma mixing between felsic rhyolitic and mafic basaltic magmas in an intermediate reservoir prior to emplacement or lava flow.

- If the erupted magma contains <52% and >45% silica, the lava is called mafic (because it contains higher percentages of magnesium (Mg) and iron (Fe)) or basaltic. These lavas are usually much less viscous than rhyolitic lavas, depending on their eruption temperature; they also tend to be hotter than felsic lavas. Mafic lavas occur in a wide range of settings:

  - ○ At mid-ocean ridges, where two oceanic plates are pulling apart, basaltic lava erupts as pillows to fill the gap,

  - ○ Shield volcanoes (e.g. the Hawaiian Islands, including Mauna Loa and Kilauea), on both oceanic and continental crust,

  - ○ As continental flood basalts.

- Some erupted magmas contain <=45% silica and produce ultramafic lava. Ultramafic flows, also known as komatiites, are very rare; indeed, very few have been erupted at the Earth's surface since the Proterozoic, when the planet's heat flow was higher. They are (or were) the hottest lavas, and probably more fluid than common mafic lavas.

Lava Texture

Two types of lava are named according to the surface texture: 'A'a , both Hawaiian words. 'A'a is characterized by a rough, clinkery surface and is the typical texture of viscous lava flows. However, even basaltic or mafic flows can be erupted as 'a'a flows, particularly if the eruption rate is high and the slope is steep.

Pāhoehoe is characterized by its smooth and often ropey or wrinkly surface and is generally formed from more fluid lava flows. Usually, only mafic flows will erupt as pāhoehoe, since they often erupt at higher temperatures or have the proper chemical make-up to allow them to flow with greater fluidity.

Volcanic Activity

Popular Classification of Volcanoes

Fresco with Mount Vesuvius behind Bacchus and
Agathodaemon, as seen in Pompeii's House of the Centenary.

A popular way of classifying magmatic volcanoes is by their frequency of eruption, with those that erupt regularly called active, those that have erupted in historical times but are now quiet called dormant or inactive, and those that have not erupted in historical times called extinct. However, these popular classifications—extinct in particular—are practically meaningless to scientists. They use classifications which refer to a particular volcano's formative and eruptive processes and resulting shapes.

Active

There is no consensus among volcanologists on how to define an "active" volcano. The lifespan of a volcano can vary from months to several million years, making such a distinction sometimes meaningless when compared to the lifespans of humans or even civilizations. For example, many of Earth's volcanoes have erupted dozens of times in the past few thousand years but are not currently showing signs of eruption. Given the long lifespan of such volcanoes, they are very active. By human lifespans, however, they are not.

Scientists usually consider a volcano to be erupting or likely to erupt if it is currently erupting, or showing signs of unrest such as unusual earthquake activity or significant new gas emissions. Most scientists consider a volcano active if it has erupted in the last 10,000 years (Holocene times)—the Smithsonian Global Volcanism Program uses this definition of active. Most volcanoes are situated on the Pacific Ring of Fire. An estimated 500 million people live near active volcanoes.

Historical time (or recorded history) is another timeframe for active. The Catalogue of the Active Volcanoes of the World, published by the International Association of Volcanology, uses this definition, by which there are more than 500 active volcanoes. However, the span of recorded history differs from region to region. In China and the Mediterranean, it reaches back nearly 3,000 years, but in the Pacific Northwest of the United States and Canada, it reaches back less than 300 years, and in Hawaii and New Zealand, only around 200 years.

Kīlauea's lava entering the sea.

Lava flows at Holuhraun, Iceland.

As of 2013, the following are considered Earth's most active volcanoes:

- Kīlauea, the famous Hawaiian volcano, has been in continuous, effusive eruption (in which lava steadily flows onto the ground) since 1983 and has the longest-observed lava lake.

- Mount Etna and nearby Stromboli, two Mediterranean volcanoes in "almost continuous eruption" since antiquity.

- Mount Yasur, in Vanuatu, has been erupting "nearly continuously" for over 800 years.

As of August 2013, the longest ongoing (but not necessarily continuous) volcanic eruptive phases are:

- Mount Yasur, 111 years,

- Mount Etna, 109 years,

- Stromboli, 108 years,

- Santa María, 101 years,

- Sangay, 94 years.

Other very active volcanoes include:

- Mount Nyiragongo and its neighbor, Nyamuragira, are Africa's most active volcanoes,

- Piton de la Fournaise, in Réunion, erupts frequently enough to be a tourist attraction,

- Erta Ale, in the Afar Triangle, has maintained a lava lake since at least 1906,

- Mount Erebus, in Antarctica, has maintained a lava lake since at least 1972,

- Mount Merapi,

- Whakaari/White Island, has been in a continuous state of releasing volcanic gas since before European observation in 1769,

- Ol Doinyo Lengai,

- Ambrym,

- Arenal Volcano,

- Pacaya,

- Klyuchevskaya Sopka,

- Sheveluch.

Nyiragongo's lava lake.

Extinct

Fourpeaked volcano, Alaska,
after being thought extinct for over 10,000 years.

Mount Rinjani eruption,
in Lombok, Indonesia.

Extinct volcanoes are those that scientists consider unlikely to erupt again because the volcano no longer has a magma supply. Examples of extinct volcanoes are many volcanoes on the Hawaiian – Emperor seamount chain in the Pacific Ocean (although some volcanoes at the eastern end of the chain are active), Hohentwiel in Germany, Shiprock in New Mexico, Zuidwal volcano in the Netherlands and many volcanoes in Italy like Monte Vulture. Edinburgh Castle in Scotland is famously located atop an extinct volcano. Otherwise, whether a volcano is truly extinct is often difficult to determine. Since "supervolcano" calderas can have eruptive lifespans sometimes measured in millions of years, a caldera that has not produced an eruption in tens of thousands of years is likely to be considered dormant instead of extinct. Some volcanologists refer to extinct volcanoes as inactive, though the term is now more commonly used for dormant volcanoes once thought to be extinct.

Dormant and Reactivated

Narcondam Island, India, is classified as a dormant
volcano by the Geological Survey of India.

It is difficult to distinguish an extinct volcano from a dormant (inactive) one. Dormant volcanoes are those that have not erupted for thousands of years, but are likely to erupt again in the future. Volcanoes are often considered to be extinct if there are no written records of its activity. Nevertheless, volcanoes may remain dormant for a long period of time. For example, Yellowstone has a repose/recharge period of around 700,000 years, and Toba of around 380,000 years. Vesuvius was described by Roman writers as having been covered with gardens and vineyards before its eruption of 79 CE, which destroyed the towns of Herculaneum and Pompeii. Before its catastrophic eruption of 1991, Pinatubo was an inconspicuous volcano, unknown to most people in the surrounding areas. Two other examples are the long-dormant Soufrière Hills volcano on the island of Montserrat, thought to be extinct before activity resumed in 1995, and Fourpeaked Mountain in Alaska, which, before its September 2006 eruption, had not erupted since before 8000 BCE and had long been thought to be extinct.

Technical Classification of Volcanoes

Volcanic-alert Level

The three common popular classifications of volcanoes can be subjective and some volcanoes thought to have been extinct have erupted again. To help prevent people from falsely believing they are not at risk when living on or near a volcano, countries have adopted new classifications to describe the various levels and stages of volcanic activity. Some alert systems use different numbers or colors to designate the different stages. Other systems use colors and words. Some systems use a combination of both.

Decade Volcanoes

The Decade Volcanoes are 16 volcanoes identified by the International Association of Volcanology and Chemistry of the Earth's Interior (IAVCEI) as being worthy of particular study in light of their history of large, destructive eruptions and proximity to populated areas. They are named Decade

Volcanoes because the project was initiated as part of the United Nations-sponsored International Decade for Natural Disaster Reduction (the 1990s). The 16 current Decade Volcanoes are:

• Avachinsky-Koryaksky (grouped together), Kamchatka, Russia.	• Sakurajima, Kagoshima Prefecture, Japan.
• Nevado de Colima, Jalisco and Colima, Mexico.	• Santa Maria/Santiaguito, Guatemala.
• Mount Etna, Sicily, Italy.	• Santorini, Cyclades, Greece.
• Galeras, Nariño, Colombia.	• Taal Volcano, Luzon, Philippines.
• Mauna Loa, Hawaii, US.	• Teide, Canary Islands, Spain.
• Mount Merapi, Central Java, Indonesia.	• Ulawun, New Britain, Papua New Guinea.
• Mount Nyiragongo, Democratic Republic of the Congo.	• Mount Unzen, Nagasaki Prefecture, Japan.
• Mount Rainier, Washington, US.	• Vesuvius, Naples, Italy.

The Deep Earth Carbon Degassing Project, an initiative of the Deep Carbon Observatory, monitors nine volcanoes, two of which are Decade volcanoes. The focus of the Deep Earth Carbon Degassing Project is to use Multi-Component Gas Analyzer System instruments to measure CO_2/SO_2 ratios in real-time and in high-resolution to allow detection of the pre-eruptive degassing of rising magmas, improving prediction of volcanic activity.

Koryaksky volcano towering over Petropavlovsk-Kamchatsky
on Kamchatka Peninsula, Far Eastern Russia.

Effects of Volcanoes

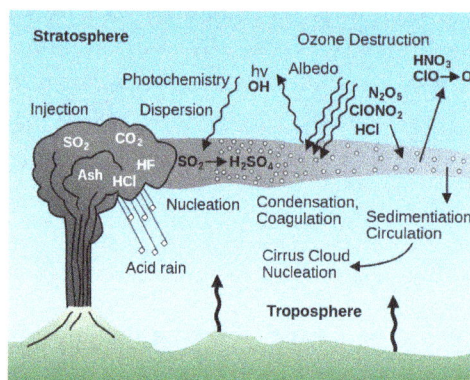

Schematic of volcano injection of aerosols and gases.

There are many different types of volcanic eruptions and associated activity: phreatic eruptions (steam-generated eruptions), explosive eruption of high-silica lava (e.g., rhyolite), effusive eruption of low-silica lava (e.g., basalt), pyroclastic flows, lahars (debris flow) and carbon dioxide emission. All of these activities can pose a hazard to humans. Earthquakes, hot springs, fumaroles, mud pots and geysers often accompany volcanic activity.

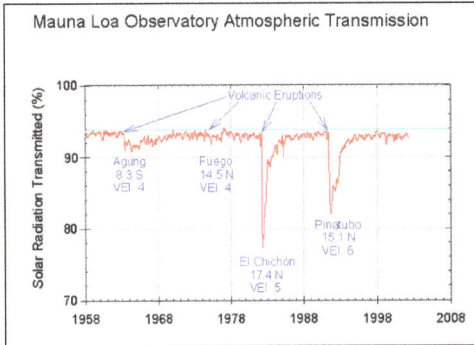

Solar radiation graph 1958–2008, showing how the radiation is reduced after major volcanic eruptions.

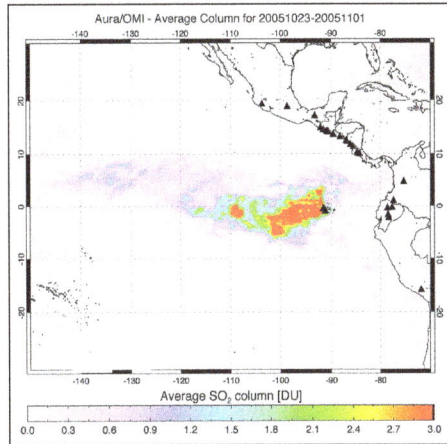

Sulfur dioxide concentration over the Sierra Negra Volcano, Galapagos Islands, during an eruption.

Volcanic Gases

The concentrations of different volcanic gases can vary considerably from one volcano to the next. Water vapor is typically the most abundant volcanic gas, followed by carbon dioxide and sulfur dioxide. Other principal volcanic gases include hydrogen sulfide, hydrogen chloride, and hydrogen fluoride. A large number of minor and trace gases are also found in volcanic emissions, for example hydrogen, carbon monoxide, halocarbons, organic compounds, and volatile metal chlorides.

Large, explosive volcanic eruptions inject water vapor (H_2O), carbon dioxide (CO_2), sulfur dioxide (SO_2), hydrogen chloride (HCl), hydrogen fluoride (HF) and ash (pulverized rock and pumice) into the stratosphere to heights of 16–32 kilometres (10–20 mi) above the Earth's surface. The most significant impacts from these injections come from the conversion of sulfur dioxide to sulfuric acid (H_2SO_4), which condenses rapidly in the stratosphere to form fine sulfate aerosols. The SO_2 emissions alone of two different eruptions are sufficient to compare their potential climatic impact. The aerosols increase the Earth's albedo—its reflection of radiation from the Sun back into space—and thus cool the Earth's lower atmosphere or troposphere; however, they also absorb heat radiated up from the Earth, thereby warming the stratosphere. Several eruptions during the past century have caused a decline in the average temperature at the Earth's surface of up to half a degree (Fahrenheit scale) for periods of one to three years; sulfur dioxide from the eruption of Huaynaputina probably caused the Russian famine of 1601–1603.

Significant Consequences

A volcanic winter is thought to have taken place around 70,000 years ago after the supereruption of Lake Toba on Sumatra island in Indonesia. According to the Toba catastrophe theory to which some anthropologists and archeologists subscribe, it had global consequences, killing most humans then alive and creating a population bottleneck that affected the genetic inheritance of all humans today.

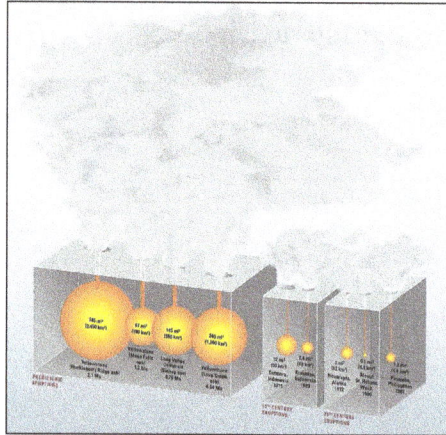

Comparison of major United States supereruptions (VEI 7 and 8) with major historical volcanic eruptions in the 19th and 20th century.

From left to right: Yellowstone 2.1 Ma, Yellowstone 1.3 Ma, Long Valley 6.26 Ma, Yellowstone 0.64 Ma. 19th century eruptions: Tambora 1815, Krakatoa 1883. 20th century eruptions: Novarupta 1912, St. Helens 1980, Pinatubo 1991.

It has been suggested that volcanic activity caused or contributed to the End-Ordovician, Permian-Triassic, Late Devonian mass extinctions, and possibly others. The massive eruptive event which formed the Siberian Traps, one of the largest known volcanic events of the last 500 million years of Earth's geological history, continued for a million years and is considered to be the likely cause of the "Great Dying" about 250 million years ago, which is estimated to have killed 90% of species existing at the time.

The 1815 eruption of Mount Tambora created global climate anomalies that became known as the "Year Without a Summer" because of the effect on North American and European weather. Agricultural crops failed and livestock died in much of the Northern Hemisphere, resulting in one of the worst famines of the 19th century.

The freezing winter of 1740–41, which led to widespread famine in northern Europe, may also owe its origins to a volcanic eruption.

Acid Rain

Sulfate aerosols promote complex chemical reactions on their surfaces that alter chlorine and nitrogen chemical species in the stratosphere. This effect, together with increased stratospheric chlorine levels from chlorofluorocarbon pollution, generates chlorine monoxide (ClO), which destroys ozone (O_3). As the aerosols grow and coagulate, they settle down into the upper troposphere where they serve as nuclei for cirrus clouds and further modify the Earth's radiation balance. Most of the hydrogen chloride (HCl) and hydrogen fluoride (HF) are dissolved in water droplets in the eruption cloud and quickly fall to the ground as acid rain. The injected ash also falls rapidly from the stratosphere; most of it is removed within several days to a few weeks. Finally, explosive volcanic eruptions release the greenhouse gas carbon dioxide and thus provide a deep source of carbon for biogeochemical cycles.

Ash plume rising from Eyjafjallajökull.

Gas emissions from volcanoes are a natural contributor to acid rain. Volcanic activity releases about 130 to 230 teragrams (145 million to 255 million short tons) of carbon dioxide each year. Volcanic eruptions may inject aerosols into the Earth's atmosphere. Large injections may cause visual effects such as unusually colorful sunsets and affect global climate mainly by cooling it. Volcanic eruptions also provide the benefit of adding nutrients to soil through the weathering process of volcanic rocks. These fertile soils assist the growth of plants and various crops. Volcanic eruptions can also create new islands, as the magma cools and solidifies upon contact with the water.

Hazards

Ash thrown into the air by eruptions can present a hazard to aircraft, especially jet aircraft where the particles can be melted by the high operating temperature; the melted particles then adhere to the turbine blades and alter their shape, disrupting the operation of the turbine. Dangerous encounters in 1982 after the eruption of Galunggung in Indonesia, and 1989 after the eruption of Mount Redoubt in Alaska raised awareness of this phenomenon. Nine Volcanic Ash Advisory Centers were established by the International Civil Aviation Organization to monitor ash clouds and advise pilots accordingly. The 2010 eruptions of Eyjafjallajökull caused major disruptions to air travel in Europe.

Bay

A bay is a recessed, coastal body of water that directly connects to a larger main body of water, such as an ocean, a lake, or another bay. A large bay is usually called a gulf, sea, sound, or bight. A cove is a type of smaller bay with a circular inlet and narrow entrance. A fjord is a particularly steep bay shaped by glacial activity.

A bay can be the estuary of a river, such as the Chesapeake Bay, an estuary of the Susquehanna River. Bays may also be nested within each other; for example, James Bay is an arm of Hudson Bay in northeastern Canada. Some large bays, such as the Bay of Bengal and Hudson Bay, have varied marine geology.

The land surrounding a bay often reduces the strength of winds and blocks waves. Bays may have as wide a variety of shoreline characteristics as other shorelines. In some cases, bays have beaches, which "are usually characterized by a steep upper foreshore with a broad, flat fronting terrace". Bays were significant in the history of human settlement because they provided safe places for fishing. Later they were important in the development of sea trade as the safe anchorage they provide encouraged their selection as ports.

The United Nations Convention on the Law of the Sea (UNCLOS) defines a bay as a well-marked indentation whose penetration is in such proportion to the width of its mouth as to contain land-locked waters and constitute more than a mere curvature of the coast. An indentation shall not, however, be regarded as a bay unless its area is as large as, or larger than, that of the semi-circle whose diameter is a line drawn across the mouth of that indentation.

Formation

There are various ways in which bays can form. The largest bays have developed through plate tectonics. As the super-continent Pangaea broke up along curved and indented fault lines, the continents moved apart and left large bays; these include the Gulf of Guinea, the Gulf of Mexico, and the Bay of Bengal, which is the world's largest bay.

Bays also form through coastal erosion by rivers and glaciers. A bay formed by a glacier is a fjord. Rias are created by rivers and are characterised by more gradual slopes. Deposits of softer rocks erode more rapidly, forming bays, while harder rocks erode less quickly, leaving headlands.

Mid-ocean Ridge

Oceanic ridge is a continuous submarine mountain chain extending approximately 80,000 km (50,000 miles) through all the world's oceans. Individually, ocean ridges are the largest features in ocean basins. Collectively, the oceanic ridge system is the most prominent feature on Earth's surface after the continents and the ocean basins themselves. In the past these features were referred to as mid-ocean ridges, but, as will be seen, the largest oceanic ridge, the East Pacific Rise, is far from a mid-ocean location, and the nomenclature is thus inaccurate. Oceanic ridges are not to be confused with aseismic ridges, which have an entirely different origin.

Principal Characteristics

Oceanic ridges are found in every ocean basin and appear to girdle Earth. The ridges rise from depths near 5 km (3 miles) to an essentially uniform depth of about 2.6 km (1.6 miles) and are roughly symmetrical in cross section. They can be thousands of kilometres wide. In places, the crests of the ridges are offset across transform faults within fracture zones, and these faults can be followed down the flanks of the ridges. (Transform faults are those along which lateral movement occurs.) The flanks are marked by sets of mountains and hills that are elongate and parallel to the ridge trend.

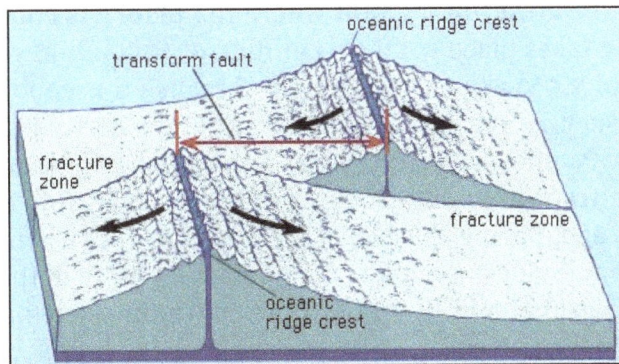

Oceanic ridges offset by transform faults and fracture zones.
The arrows show the direction of movement across the transform faults.

New oceanic crust (and part of Earth's upper mantle, which, together with the crust, makes up the lithosphere) is formed at seafloor spreading centres at these crests of the oceanic ridges. Because of this, certain unique geologic features are found there. Fresh basaltic lavas are exposed on the seafloor at the ridge crests. These lavas are progressively buried by sediments as the seafloor spreads away from the site. The flow of heat out of the crust is many times greater at the crests than elsewhere in the world. Earthquakes are common along the crests and in the transform faults that join the offset ridge segments. Analysis of earthquakes occurring at the ridge crests indicates that the oceanic crust is under tension there. A high-amplitude magnetic anomaly is centred over the crests because fresh lavas at the crests are being magnetized in the direction of the present geomagnetic field.

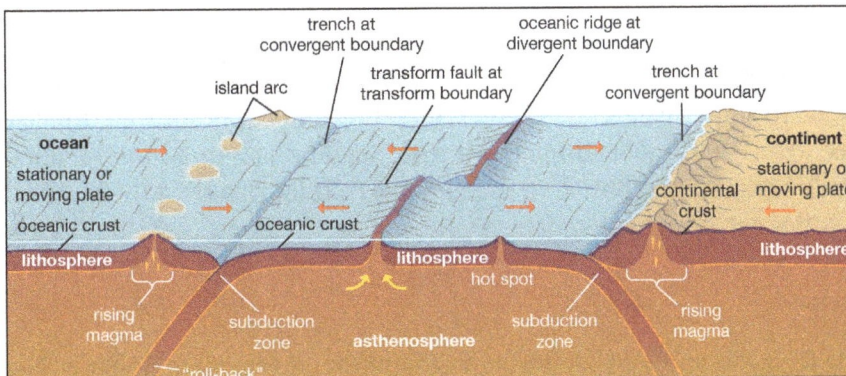

Crustal generation and destruction.

Three-dimensional diagram showing crustal generation and destruction according to the theory of plate tectonics; included are the three kinds of plate boundaries—divergent, convergent (or collision), and strike-slip (or transform).

The depths over the oceanic ridges are rather precisely correlated with the age of the ocean crust; specifically, it has been demonstrated that the ocean depth is proportional to the square root of crustal age. The theory explaining this relationship holds that the increase in depth with age is due to the thermal contraction of the oceanic crust and upper mantle as they are carried away from the seafloor spreading centre in an oceanic plate. Because such a tectonic plate is ultimately about 100 km (62 miles) thick, contraction of only a few percent predicts the entire relief of an oceanic ridge. It then follows that the width of a ridge can be defined as twice the distance from the crest to the point where the plate has cooled to a steady thermal state. Most of the cooling takes place within 70 million or 80 million years, by which time the ocean depth is about 5 to 5.5 km (3.1 to 3.5 miles). Because this cooling is a function of age, slow-spreading ridges, such as the Mid-Atlantic Ridge, are narrower than faster-spreading ridges, such as the East Pacific Rise. Further, a correlation has been found between global spreading rates and the transgression and regression of ocean waters onto the continents. About 100 million years ago, during the early Cretaceous Period when global spreading rates were uniformly high, oceanic ridges occupied comparatively more of the ocean basins, causing the ocean waters to transgress (spill over) onto the continents, leaving marine sediments in areas now well away from coastlines.

Besides ridge width, other features appear to be a function of spreading rate. Global spreading rates range from 10 mm (0.4 inch) per year or less up to 160 mm (6.3 inches) per year. Oceanic ridges can be classified as slow (up to 50 mm [about 2 inches] per year), intermediate (up to 90 mm [about 3.5 inches] per year), and fast (up to 160 mm per year). Slow-spreading ridges are characterized by a rift valley at the crest. Such a valley is fault-controlled. It is typically 1.4 km (0.9 mile) deep and 20–40 km (about 12–25 miles) wide. Faster-spreading ridges lack rift valleys. At intermediate rates, the crest regions are broad highs with occasional fault-bounded valleys no deeper than 200 metres (about 660 feet). At fast rates, an axial high is present at the crest. The slow-spreading rifted ridges have rough faulted topography on their flanks, while the faster-spreading ridges have much smoother flanks.

Distribution of Major Ridges and Spreading Centres

Oceanic spreading centres are found in all the ocean basins. In the Arctic Ocean a slow-rate spreading centre is located near the eastern side in the Eurasian basin. It can be followed south, offset by transform faults, to Iceland. Iceland has been created by a hot spot located directly below an oceanic spreading centre. The ridge leading south from Iceland is named the Reykjanes Ridge, and, although it spreads at 20 mm (0.8 inch) per year or less, it lacks a rift valley. This is thought to be the result of the influence of the hot spot.

Atlantic Ocean

The Mid-Atlantic Ridge extends from south of Iceland to the extreme South Atlantic Ocean near 60° S latitude. It bisects the Atlantic Ocean basin, which led to the earlier designation of mid-ocean ridge for features of this type. The Mid-Atlantic Ridge became known in a rudimentary fashion during the 19th century. In 1855 Matthew Fontaine Maury of the U.S. Navy prepared a chart of the Atlantic in which he identified it as a shallow "middle ground." During the 1950s the American oceanographers Bruce Heezen and Maurice Ewing proposed that it was a continuous mountain range.

In the North Atlantic the ridge spreads slowly and displays a rift valley and mountainous flanks. In the South Atlantic spreading rates are between slow and intermediate, and rift valleys are generally absent, as they occur only near transform faults.

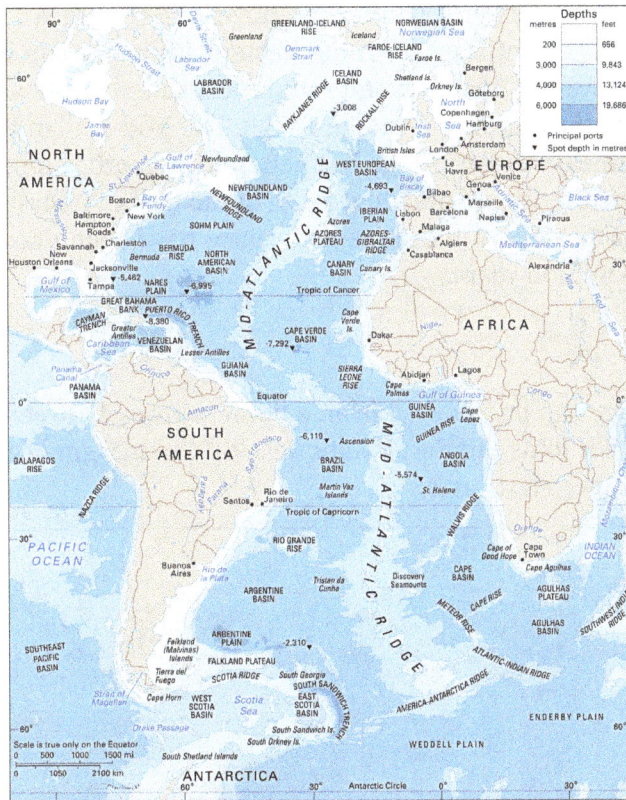

The Atlantic Ocean, with depth contours and submarine features.

Indian Ocean

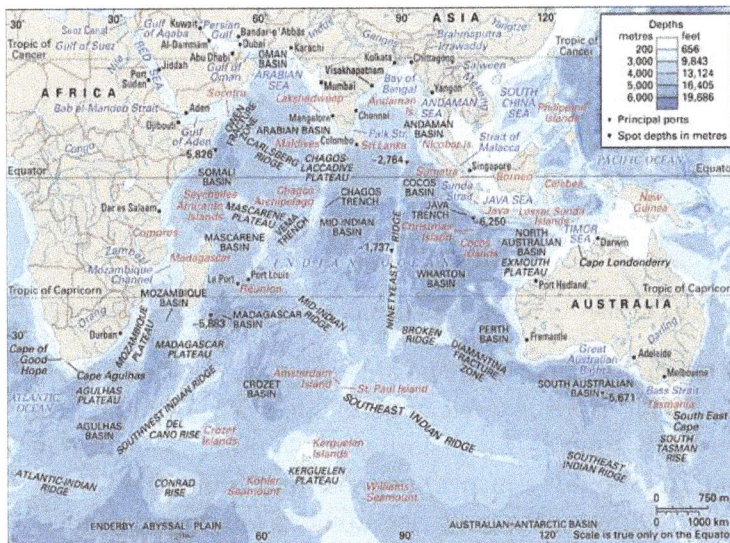

Indian ocean.

A very slow oceanic ridge, the Southwest Indian Ridge, bisects the ocean between Africa and Antarctica. It joins the Mid-Indian and Southeast Indian ridges east of Madagascar. The Carlsberg Ridge is found at the north end of the Mid-Indian Ridge. It continues north to join spreading centres in the Gulf of Aden and Red Sea. Spreading is very slow at this point but approaches intermediate rates on the Carlsberg and Mid-Indian ridges. The Southeast Indian Ridge spreads at intermediate rates. This ridge continues from the western Indian Ocean in a southeasterly direction, bisecting the ocean between Australia and Antarctica. Rifted crests and rugged mountainous flanks are characteristic of the Southwest Indian Ridge. The Mid-Indian Ridge has fewer features of this kind, and the Southeast Indian Ridge has generally smoother topography. The latter also displays distinct asymmetric seafloor spreading south of Australia. Analysis of magnetic anomalies shows that rates on opposite sides of the spreading centre have been unequal at many times over the past 50 million or 60 million years.

Pacific Ocean

The Pacific-Antarctic Ridge can be followed from a point midway between New Zealand and Antarctica northeast to where it joins the East Pacific Rise off the margin of South America. The former spreads at intermediate to fast rates.

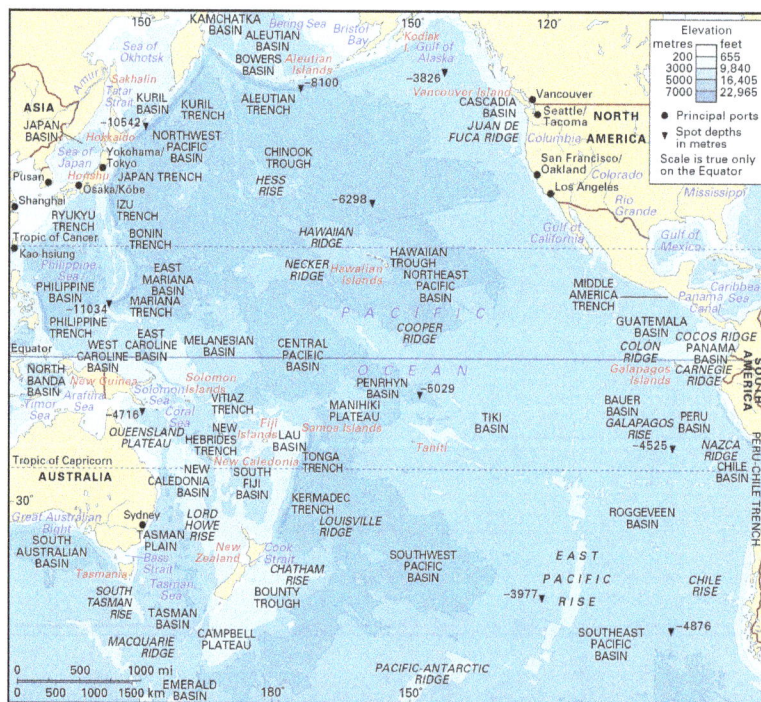

The Pacific Ocean, with depth contours and submarine features.

The East Pacific Rise extends from this site northward to the Gulf of California, where it joins the transform zone of the Pacific-North American plate boundary. Offshore from Chile and Peru, the East Pacific Rise is currently spreading at fast rates of 159 mm (6.3 inches) per year or more. Rates decrease to about 60 mm (about 2.4 inches) per year at the mouth of the Gulf of California. The crest of the ridge displays a low topographic rise along its length rather than a rift valley. The East Pacific Rise was first detected during the Challenger Expedition of the 1870s. It was described in its gross form during the 1950s and '60s by oceanographers, including Heezen, Ewing, and Henry W. Menard. During the 1980s, Kenneth C. Macdonald, Paul J. Fox, and Peter F. Lonsdale discovered that the main spreading centre appears to be interrupted and offset a few kilometres to one side at various places along the crest of the East Pacific Rise. However, the ends of the offset spreading centres overlap each other by several kilometres. These were identified as a new type of geologic feature of oceanic spreading centres and were designated overlapping spreading centres. Such centres are thought to result from interruptions of the magma supply to the crest along its length and define a fundamental segmentation of the ridge on a scale of tens to hundreds of kilometres.

Many smaller spreading centres branch off the major ones or are found behind island arcs. In the western Pacific, spreading centres occur on the Fiji Plateau between the New Hebrides and Fiji Islands and in the Woodlark Basin between New Guinea and the Solomon Islands. A series of spreading centres and transform faults lie between the East Pacific Rise and South America near 40° to 50° S latitude. The Scotia Sea between South America and the Antarctic Peninsula contains a spreading centre. The Galapagos spreading centre trends east-west between the East Pacific Rise and South America near the Equator. Three short spreading centres are found a few hundred kilometres off the shore of the Pacific Northwest. These are the Gorda Ridges off northern California, the Juan de Fuca Ridge off Oregon and Washington, and the Explorer Ridge off Vancouver Island.

In a careful study of the seafloor spreading history of the Galapagos and the Juan de Fuca spreading centres, the American geophysicist Richard N. Hey developed the idea of the propagating rift. In this phenomenon, one branch of a spreading centre ending in a transform fault lengthens at the expense of the spreading centre across the fault. The rift and fault propagate at one to five times the spreading rate and create chevron patterns in magnetic anomalies and the grain of the seafloor topography resembling the wake of a boat.

Spreading Centre Zones and Associated Phenomena

From the 1970s highly detailed studies of spreading centres using deeply towed instruments, photography, and manned submersibles have resulted in new revelations about the processes of seafloor spreading. The most profound discoveries have been of deep-sea hydrothermal vents and previously unknown biological communities.

Spreading Centre Zones

Spreading centres are divided into several geologic zones. The neovolcanic zone is at the very axis. It is 1–2 km (0.6–1.2 miles) wide and is the site of recent and active volcanism and of the hydrothermal vents. It is marked by chains of small volcanoes or volcanic ridges. Adjacent to the neovolcanic zone is one marked by fissures in the seafloor. This may be 1 to 2 km wide. Beyond this point occurs a zone of active faulting. Here, fissures develop into normal faults with vertical offsets. This zone may be 10 km (about 6 miles) or more wide. At slow spreading rates the faults have offsets of hundreds of metres, creating rift valleys and rift mountains. At faster rates the vertical offsets are 50 metres (about 160 feet) or less. A deep rift valley is not formed because the vertical uplifts are cancelled out by faults that downdrop uplifted blocks. This results in linear, fault-bounded abyssal hills and valleys trending parallel to the spreading centre.

Warm springs emanating from the seafloor in the neovolcanic zone were first found on the Galapagos spreading centre. These waters were measured to have temperatures about 20 °C (36 °F) above the ambient temperature. In 1979 hydrothermal vents with temperatures near 350 °C (662 °F) were discovered on the East Pacific Rise off Mexico. Since then similar vents have been found on the spreading centres off the Pacific Northwest coast of the United States, on the south end of the northern Mid-Atlantic Ridge, and at many locations on the East Pacific Rise.

Hydrothermal Vents

Hydrothermal vents are localized discharges of heated seawater. They result from cold seawater percolating down into the hot oceanic crust through the zone of fissures and returning to the seafloor in a pipelike flow at the axis of the neovolcanic zone. The heated waters often carry sulfide minerals of zinc, iron, and copper leached from the crust. Outflow of these heated waters probably accounts for 20 percent of Earth's heat loss. Exotic biological communities exist around the hydrothermal vents. These ecosystems are totally independent of energy from the Sun. They are not dependent on photosynthesis but rather on chemosynthesis by sulfur-fixing bacteria. The sulfide minerals precipitated in the neovolcanic zone can accumulate in substantial amounts and are sometimes buried by lava flows at a later time. Such deposits are mined as commercial ores in ophiolites on Cyprus and in Oman.

Magma Chambers

Magma chambers have been detected beneath the crest of the East Pacific Rise by seismic experiments. (The principle underlying the experiments is that partially molten or molten rock slows the travel of seismic waves and also strongly reflects them.) The depth to the top of the chambers is about 2 km (1.2 miles) below the seafloor. The width is more difficult to ascertain but is probably 1 to 4 km (0.6 to 2.5 miles). Their thickness seems to be about 2 to 6 km (1.2 to 3.7 miles), on the basis of studies of ophiolites. The chambers have been mapped along the trend of the crest between 9° and 13° N latitude. The top is relatively continuous, but is apparently interrupted by offsets of transform faults and overlapping spreading centres.

References

- Continental-landform, science: britannica.com, Retrieved 18 January, 2019

- Fraknoi, A.; Morrison, D.; Wolff, S. (2004). Voyages to the Planets (3rd ed.). Belmont: Thomson Books/Cole. ISBN 978-0-534-39567-4

- Jørgensen, Flemming; Peter B.E. Sandersen (June 2006). "Buried and open tunnel valleys in Denmark—erosion beneath multiple ice sheets". Quaternary Science Reviews. 25 (11–12): 1339–1363. Bibcode:2006QSRv...25.1339J. doi:10.1016/j.quascirev.2005.11.006

- oceanic-ridge, science: britannica.com, Retrieved 23 July, 2019

- Rood, Stewart B.; Pan, Jason; Gill, Karen M.; Franks, Carmen G.; Samuelson, Glenda M.; Shepherd, Anita (2008-02-01). "Declining summer flows of Rocky Mountain rivers: Changing seasonal hydrology and probable impacts on floodplain forests". Journal of Hydrology. 349 (3–4): 397–410. doi:10.1016/j.jhydrol.2007.11.012

- Garcia-Castellanos, D., 2007. The role of climate during high plateau formation. Insights from numerical experiments. Earth Planet. Sci. Lett. 257, 372-390, doi:10.1016/j.epsl.2007.02.039 pdf

3
Different Approaches to Study Topography

There are a number of ways in which topography can be studied. These include aerial survey, land surveying, compass surveying, remote sensing, photogrammetry, satellite imagery, digital elevation model, geovisualization, hypsometric tinting, etc. These diverse approaches for studying topography in the current scenario have been thoroughly discussed in this chapter.

Aerial Survey

Aerial Survey is a form of collection of geographical information using airborne vehicles. The collection of information can be made using different technologies such as aerial photography, radar, laser or from remote sensing imagery using other bands of the electromagnetic spectrum, such as infrared, gamma, or ultraviolet. For the information collected to be useful this information needs to be georeferenced. The georeferencing of information is usually done using GNSS with similar techniques as the techniques used for dynamic land surveying.

Application Architecture

Aerial Survey Plane.

Aerial Surveying is normally done using manned aeroplanes where the sensors (cameras, radars, lasers, detectors, etc) and the GNSS receiver are setup and are calibrated for the adequate georeferencing of the collected data. Apart from manned aeroplanes, other aerial vehicles can be also used such as UAVs, balloons, helicopters.

Usually for this type of applications, kinematic methods are used. The algorithms used in these dynamic surveys rely on the fact that while the receiver moves through the air should never loose lock on the satellites signal. The techniques and algorithms used can be used in post-processing since usually the positioning data is not required in real-time.

Multiple sensors of different or similar types can be used in order to collect different types of information or to be able to build 3D computer models of the terrain (e.g. stereoscopic cameras).

The data collected can be used for different purposes such as:

- Sea surveys (sea level, temperature, undulation, etc.),
- Land survey (cartography, topography, feature recognition, etc.),
- Monitoring vegetation and ground cover,
- Reconnaissance.

Application Characterization

The 3D visualization of the data collected by aerial surveys can be created by georeferencing the aerial photos and other sensor data (laser, radar, etc.) in the same reference frame, orthorectifying the aerial photos, and then draping the orthorectified images on top of the other sensors grid. It is also possible to create digital terrain models and thus 3D visualisations using multiple aerial photographs. Techniques such as adaptive least squares stereo matching are then used to produce a dense array of correspondences which are transformed through a camera model to produce a dense array of x, y, z data which can be used to produce digital terrain model and orthoimage products.

Application Examples

Two of the most used forms of Aerial Survey are:

- Aerial Laser Profiling - Aerial Laser Profiling uses short duration laser pulses that are emitted towards the ground, reflected and detected by a receiver in the airborne vehicle. The time between the emission of the pulse and the reception can be used to determine the distance travelled by the pulse.

- Aerial Photogrammetry - In aerial photogrammetry aerial photos are taken in order to produce 2D or 3D terrain models. Multiple cameras might need to be used to build 3D models.

Land Surveying

Surveying or land surveying is the technique, profession, art and science of determining the terrestrial or three-dimensional positions of points and the distances and angles between them. A land surveying professional is called a land surveyor. These points are usually on the surface of the Earth, and they are often used to establish maps and boundaries for ownership, locations, such as building corners or the surface location of subsurface features, or other purposes required by government or civil law, such as property sales.

Surveyors work with elements of geometry, trigonometry, regression analysis, physics, engineering, metrology, programming languages, and the law. They use equipment, such as total stations, robotic total stations, theodolites, GNSS receivers, retroreflectors, 3D scanners, radios, clinometer, handheld tablets, digital levels, subsurface locators, drones, GIS, and surveying software.

Surveying has been an element in the development of the human environment since the beginning of recorded history. The planning and execution of most forms of construction require it. It is also used in transport, communications, mapping, and the definition of legal boundaries for land ownership. It is an important tool for research in many other scientific disciplines.

The International Federation of Surveyors defines the function of surveying.

A surveyor is a professional person with the academic qualifications and technical expertise to conduct one, or more, of the following activities:

- To determine, measure and represent land, three-dimensional objects, point-fields and trajectories,

- To assemble and interpret land and geographically related information,

- To use that information for the planning and efficient administration of the land, the sea and any structures thereon,

- To conduct research into the above practices and to develop them.

Equipment

Hardware

The main surveying instruments in use around the world are the theodolite, measuring tape, total station, 3D scanners, GPS/GNSS, level and rod. Most instruments screw onto a tripod when in use. Tape measures are often used for measurement of smaller distances. 3D scanners and various forms of aerial imagery are also used.

The theodolite is an instrument for the measurement of angles. It uses two separate *circles, protractors* or *alidades* to measure angles in the horizontal and the vertical plane. A telescope mounted on trunnions is aligned vertically with the target object. The whole upper section rotates for horizontal alignment. The vertical circle measures the angle that the telescope makes against the vertical, known as the zenith angle. The horizontal circle uses an upper and lower plate. When beginning the survey, the surveyor points the instrument in a known direction (bearing), and clamps the lower plate in place. The instrument can then rotate to measure the bearing to other objects. If no bearing is known or direct angle measurement is wanted, the instrument can be set to zero during the initial sight. It will then read the angle between the initial object, the theodolite itself, and the item that the telescope aligns with.

The gyrotheodolite is a form of theodolite that uses a gyroscope to orient itself in the absence of reference marks. It is used in underground applications.

The total station is a development of the theodolite with an electronic distance measurement device (EDM). A total station can be used for leveling when set to the horizontal plane. Since their introduction, total stations have shifted from optical-mechanical to fully electronic devices.

Surveying equipment. Clockwise from upper left: optical theodolite, robotic total station, RTK GPS base station, optical level.

Modern top-of-the-line total stations no longer need a reflector or prism to return the light pulses used for distance measurements. They are fully robotic, and can even e-mail point data to a remote computer and connect to satellite positioning systems, such as Global Positioning System. Real Time Kinematic GPS systems have increased the speed of surveying, but they are still only horizontally accurate to about 20 mm and vertically to 30–40 mm.

GPS surveying differs from other GPS uses in the equipment and methods used. Static GPS uses two receivers placed in position for a considerable length of time. The long span of time lets the receiver compare measurements as the satellites orbit. The changes as the satellites orbit also provide the measurement network with well-conditioned geometry. This produces an accurate baseline that can be over 20 km long. RTK surveying uses one static antenna and one roving antenna. The static antenna tracks changes in the satellite positions and atmospheric conditions. The surveyor uses the roving antenna to measure the points needed for the survey. The two antennas use a radio link that allows the static antenna to send corrections to the roving antenna. The roving antenna then applies those corrections to the GPS signals it is receiving to calculate its own position. RTK surveying covers smaller distances than static methods. This is because divergent conditions further away from the base reduce accuracy.

Surveying instruments have characteristics that make them suitable for certain uses. Theodolites and levels are often used by constructors rather than surveyors in first world countries. The constructor can perform simple survey tasks using a relatively cheap instrument. Total stations are workhorses for many professional surveyors because they are versatile and reliable in all conditions. The productivity improvements from a GPS on large scale surveys makes them popular for major infrastructure or data gathering projects. One-person robotic-guided total stations allow surveyors to measure without extra workers to aim the telescope or record data. A fast but expensive way to measure large areas is with a helicopter, using a GPS to record the location of the helicopter and a laser scanner to measure the ground. To increase precision, surveyors place beacons on the ground (about 20 km (12 mi) apart). This method reaches precisions between 5–40 cm (depending on flight height).

Surveyors use ancillary equipment such as tripods and instrument stands; staves and beacons used for sighting purposes; PPE; vegetation clearing equipment; digging implements for finding survey markers buried over time; hammers for placements of markers in various surfaces and structures; and portable radios for communication over long lines of sight.

Software

Land surveyors, construction professionals and civil engineers using total station, GPS, 3D scanners and other collector data use Land Surveying Software to increase efficiency, accuracy and productivity. Land Surveying Software is a staple of contemporary land surveying.

Techniques

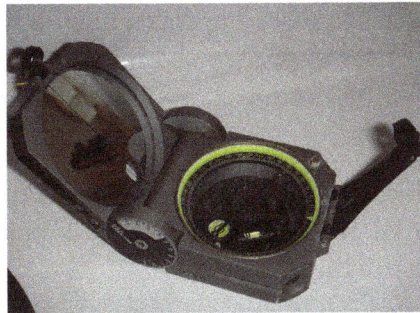

A standard Brunton Geo compass, still used commonly today by geographers, geologists and surveyors for field-based measurements.

Surveyors determine the position of objects by measuring angles and distances. The factors that can affect the accuracy of their observations are also measured. They then use this data to create vectors, bearings, coordinates, elevations, areas, volumes, plans and maps. Measurements are often split into horizontal and vertical components to simplify calculation. GPS and astronomic measurements also need measurement of a time component.

Distance Measurement

Before EDM devices, distances were measured using a variety of means. These included chains with links of a known length such as a Gunter's chain, or measuring tapes made of steel or invar. To measure horizontal distances, these chains or tapes were pulled taut to reduce sagging and slack. The distance had to be adjusted for heat expansion. Attempts to hold the measuring instrument level would also be made. When measuring up a slope, the surveyor might have to "break" (break chain) the measurement- use an increment less than the total length of the chain. Perambulators, or measuring wheels, were used to measure longer distances but not to a high level of accuracy. Tacheometry is the science of measuring distances by measuring the angle between two ends of an object with a known size. It was sometimes used before to the invention of EDM where rough ground made chain measurement impractical.

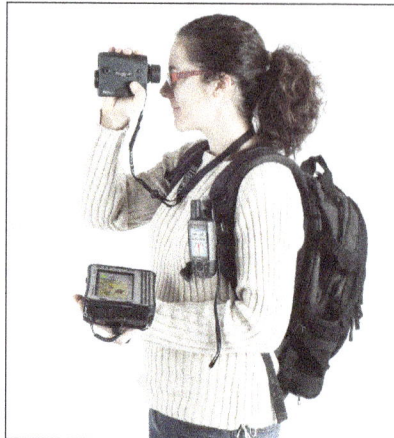

Example of modern equipment for surveying (Field-Map technology): GPS, laser rangefinder and field computer allows surveying as well as cartography (creation of map in real-time) and field data collection.

Angle Measurement

Historically, horizontal angles were measured by using a compass to provide a magnetic bearing or azimuth. Later, more precise scribed discs improved angular resolution. Mounting telescopes with reticles atop the disc allowed more precise sighting. Levels and calibrated circles allowed measurement of vertical angles. Verniers allowed measurement to a fraction of a degree, such as with a turn-of-the-century transit.

The plane table provided a graphical method of recording and measuring angles, which reduced the amount of mathematics required. In 1829 Francis Ronalds invented a reflecting instrument for recording angles graphically by modifying the octant.

By observing the bearing from every vertex in a figure, a surveyor can measure around the figure. The final observation will be between the two points first observed, except with a 180° difference. This is called a *close*. If the first and last bearings are different, this shows the error in the survey, called the *angular misclose*. The surveyor can use this information to prove that the work meets the expected standards.

Levelling

Center for Operational Oceanographic Products and Services staff member conducts tide station levelling in support of the US Army Corp of Engineers.

The simplest method for measuring height is with an altimeter using air pressure to find height. When more precise measurements are needed, means like precise levels (also known as differential leveling) are used. When precise leveling, a series of measurements between two points are taken using an instrument and a measuring rod. Differences in height between the measurements are added and subtracted in a series to get the net difference in elevation between the two endpoints. With the Global Positioning System (GPS), elevation can be measured with satellite receivers. Usually GPS is somewhat less accurate than traditional precise leveling, but may be similar over long distances.

When using an optical level, the endpoint may be out of the effective range of the instrument. There may be obstructions or large changes of elevation between the endpoints. In these situations, extra setups are needed. *Turning* is a term used when referring to moving the level to take an elevation shot from a different location. To "turn" the level, one must first take a reading and record the elevation of the point the rod is located on. While the rod is being kept in exactly the same location, the level is moved to a new location where the rod is still visible. A reading is taken from the new location of the level and the height difference is used to find the new elevation of the level gun. This is repeated until the series of measurements is completed. The level must be horizontal to get a valid measurement. Because of this, if the horizontal crosshair of the instrument is lower than the base of the rod, the surveyor will not be able to sight the rod and get a reading. The rod can usually be raised up to 25 feet high, allowing the level to be set much higher than the base of the rod.

Determining Position

The primary way of determining one's position on the earth's surface when no known positions are nearby is by astronomic observations. Observations to the sun, moon and stars could all be made using navigational techniques. Once the instrument's position and bearing to a star is determined, the bearing can be transferred to a reference point on the earth. The point can then be used as a base for further observations. Survey-accurate astronomic positions were difficult to observe and calculate and so tended to be a base off which many other measurements were made. Since the advent of the GPS system, astronomic observations are rare as GPS allows adequate positions to be determined over most of the surface of the earth.

Reference Networks

Few survey positions are derived from first principles. Instead, most surveys points are measured relative to previous measured points. This forms a reference or *control* network where each point can be used by a surveyor to determine their own position when beginning a new survey.

Survey points are usually marked on the earth's surface by objects ranging from small nails driven into the ground to large beacons that can be seen from long distances. The surveyors can set up their instruments on this position and measure to nearby objects. Sometimes a tall, distinctive feature such as a steeple or radio aerial has its position calculated as a reference point that angles can be measured against.

Triangulation is a method of horizontal location favoured in the days before EDM and GPS measurement. It can determine distances, elevations and directions between distant objects. Since the early days of surveying, this was the primary method of determining accurate positions of objects for topographic maps of large areas. A surveyor first needs to know the horizontal distance between two of the objects, known as the baseline. Then the heights, distances and angular position of other objects can be derived, as long as they are visible from one of the original objects. High-accuracy transits or theodolites were used, and angle measurements repeated for increased accuracy.

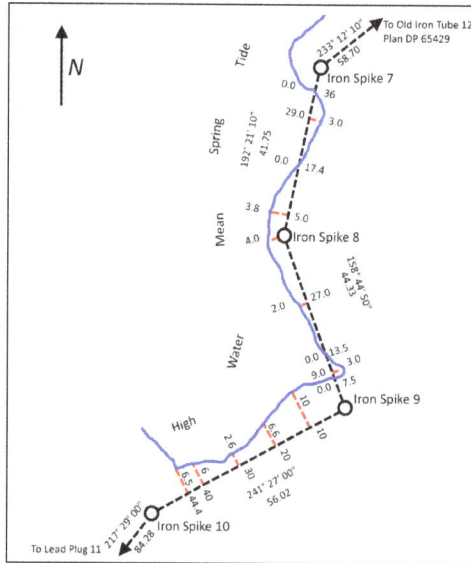

A survey using traverse and offset measurements to record the location of the shoreline shown in blue. Black dashed lines are traverse measurements between reference points (black circles). The red lines are offsets measured at right angles to the traverse lines.

Offsetting is an alternate method of determining position of objects, and was often used to measure imprecise features such as riverbanks. The surveyor would mark and measure two known positions on the ground roughly parallel to the feature, and mark out a baseline between them. At regular intervals, a distance was measured at right angles from the first line to the feature. The measurements could then be plotted on a plan or map, and the points at the ends of the offset lines could be joined to show the feature.

Traversing is a common method of surveying smaller areas. The surveyor starts from an old reference mark or known position and places a network of reference marks covering the survey area. They then measure bearings and distances between the reference marks, and to the target features. Most traverses form a loop pattern or link between two prior reference marks so the surveyor can check their measurements.

Datum and Coordinate Systems

Many surveys do not calculate positions on the surface of the earth, but instead measure the relative positions of objects. However, often the surveyed items need to be compared to outside data, such as boundary lines or previous survey's objects. The oldest way of describing a position is via latitude and longitude, and often a height above sea level. As the surveying profession grew it created Cartesian coordinate systems to simplify the mathematics for surveys over small parts of the earth. The simplest coordinate systems assume that the earth is flat and measure from an arbitrary point, known as a 'datum' (singular form of data). The coordinate system allows easy calculation of the distances and direction between objects over small areas. Large areas distort due to the earth's curvature. North is often defined as true north at the datum.

For larger regions, it is necessary to model the shape of the earth using an ellipsoid or a geoid. Many countries have created coordinate-grids customized to lessen error in their area of the earth.

Errors and Accuracy

A basic tenet of surveying is that no measurement is perfect, and that there will always be a small amount of error. There are three classes of survey errors:

- Gross errors or blunders: Errors made by the surveyor during the survey. Upsetting the instrument, misaiming a target, or writing down a wrong measurement are all gross errors. A large gross error may reduce the accuracy to an unacceptable level. Therefore, surveyors use redundant measurements and independent checks to detect these errors early in the survey.

- Systematic: Errors that follow a consistent pattern. Examples include effects of temperature on a chain or EDM measurement, or a poorly adjusted spirit-level causing a tilted instrument or target pole. Systematic errors that have known effects can be compensated or corrected.

- Random: Random errors are small unavoidable fluctuations. They are caused by imperfections in measuring equipment, eyesight, and conditions. They can be minimized by redundancy of measurement and avoiding unstable conditions. Random errors tend to cancel each other out, but checks must be made to ensure they are not propagating from one measurement to the next.

Surveyors avoid these errors by calibrating their equipment, using consistent methods, and by good design of their reference network. Repeated measurements can be averaged and any outlier measurements discarded. Independent checks like measuring a point from two or more locations or using two different methods are used. Errors can be detected by comparing the results of the two measurements.

Once the surveyor has calculated the level of the errors in his or her work, it is adjusted. This is the process of distributing the error between all measurements. Each observation is weighted according to how much of the total error it is likely to have caused and part of that error is allocated to it in a proportional way. The most common methods of adjustment are the Bowditch method, also known as the compass rule, and the principle of least squares method.

The surveyor must be able to distinguish between accuracy and precision. In the United States, surveyors and civil engineers use units of feet wherein a survey foot breaks down into 10ths and 100ths. Many deed descriptions containing distances are often expressed using these units (125.25 ft). On the subject of accuracy, surveyors are often held to a standard of one one-hundredth of a foot; about 1/8 inch. Calculation and mapping tolerances are much smaller wherein achieving near-perfect closures are desired. Though tolerances will vary from project to project, in the field and day to day usage beyond a 100th of a foot is often impractical.

Types

Local organisations or regulatory bodies class specializations of surveying in different ways. Broad groups are:

- As-built survey: A survey that documents the location of recently constructed elements of a construction project. As-built surveys are done for record, completion evaluation and payment purposes. An as-built survey is also known as a 'works as executed survey'. As-built surveys are often presented in red or redline and laid over existing plans for comparison with design information.

- Cadastral or boundary surveying: A survey that establishes or re-establishes boundaries of a parcel using a legal description. It involves the setting or restoration of monuments or markers at the corners or along the lines of the parcel. These take the form of iron rods, pipes, or concrete monuments in the ground, or nails set in concrete or asphalt. The ALTA/ACSM Land Title Survey is a standard proposed by the American Land Title Association and the American Congress on Surveying and Mapping. It incorporates elements of the boundary survey, mortgage survey, and topographic survey.

- Control surveying: Control surveys establish reference points to use as starting positions for future surveys. Most other forms of surveying will contain elements of control surveying.

- Construction surveying.

- Deformation survey: A survey to determine if a structure or object is changing shape or moving. First the positions of points on an object are found. A period of time is allowed to pass and the positions are then re-measured and calculated. Then a comparison between the two sets of positions is made.

- Dimensional control survey: This is a type of survey conducted in or on a non-level surface. Common in the oil and gas industry to replace old or damaged pipes on a like-for-like basis, the advantage of dimensional control survey is that the instrument used to conduct the survey does not need to be level. This is useful in the off-shore industry, as not all platforms are fixed and are thus subject to movement.

- Engineering surveying: Topographic, layout, and as-built surveys associated with engineering design. They often need geodetic computations beyond normal civil engineering practice.

- Foundation survey: A Survey done to collect the positional data on a foundation that has been poured and is cured. This is done to ensure that the foundation was constructed in the location, and at the elevation, authorized in the plot plan, site plan, or subdivision plan.

- Hydrographic survey: A survey conducted with the purpose of mapping the shoreline and bed of a body of water. Used for navigation, engineering, or resource management purposes.

- Leveling: Either finds the elevation of a given point or establish a point at a given elevation.

- LOMA survey: Survey to change base flood line, removing property from a SFHA special flood hazard area.

- Measured survey: A building survey to produce plans of the building. Such a survey may be conducted before renovation works, for commercial purpose, or at end of the construction process.

- Mining surveying: Mining surveying includes directing the digging of mine shafts and galleries and the calculation of volume of rock. It uses specialised techniques due to the restraints to survey geometry such as vertical shafts and narrow passages.

- Mortgage survey: A mortgage survey or physical survey is a simple survey that delineates land boundaries and building locations. It checks for encroachment, building setback restrictions and shows nearby flood zones. In many places a mortgage survey is a precondition for a mortgage loan.

- Photographic control survey: A survey that creates reference marks visible from the air to allow aerial photographs to be rectified.

- Stakeout, layout or setout: An element of many other surveys where the calculated or proposed position of an object is marked on the ground. This can be temporary or permanent. This is an important component of engineering and cadastral surveying.

- Structural survey: A detailed inspection to report upon the physical condition and structural stability of a building or structure. It highlights any work needed to maintain it in good repair.

- Subdivision: A boundary survey that splits a property into two or more smaller properties.

- Topographic survey: A survey that measures the elevation of points on a particular piece of land, and presents them as contour lines on a plot.

Plane and Geodetic Surveying

Based on the considerations and true shape of the earth, surveying is broadly classified into two types:

- Plane surveying assumes the earth is flat. Curvature and spheroidal shape of the earth is neglected. In this type of surveying all triangles formed by joining survey lines are considered as plane triangles. It is employed for small survey works where errors due to the earth's shape are too small to matter.

- In geodetic surveying the curvature of the earth is taken into account while calculating reduced levels, angles, bearings and distances. This type of surveying is usually employed for large survey works. Survey works up to 100 square miles (260 square kilometers) are treated as plane and beyond that are treated as geodetic. In geodetic surveying necessary corrections are applied to reduced levels, bearings and other observations.

Compass Surveying

Compass surveying is the branch of surveying in which the position of an object is located using angular measurements determined by a compass and linear measurements using a chain or tape. Compass surveying is used in following circumstances:

- If the surveying area is large, chain surveying is not adopted for surveying rather compass surveying is employed.

- If the plot for surveying has numerous obstacles and undulations which prevents chaining.

- If there is a time limit for surveying, compass surveying is usually adopted.

Prismatic compass.

Compass surveying is not used in places which contain iron core, power lines etc which usually attracts magnets due to their natural properties and electromagnetic properties respectively. Compass surveying is done by using traversing. A traverse is formed by connecting the points in the plot by means of a series of straight lines.

Magnetic Compass

Magnetic compass is used to find out the magnetic bearing of survey lines. The bearings may either measured in Whole Circle Bearing (W.C.B) system or in Quadrantal Bearing (Q.B) system based on the type of compass used. The basic principle of magnetic compass is if a strip of steel or iron is magnetized and pivoted exactly at centre so that it can swing freely, then it will establish itself in the magnetic meridian at the place of arrangement.

Major types of magnetic compass are:

- Prismatic compass.

- Surveyor's compass.

- Level compass.

Prismatic Compass

Prismatic compass is a portable magnetic compass which can be either used as a hand instrument or can be fitted on a tripod. It contains a prism which is used for accurate measurement of readings. The greatest advantage of this compass is both sighting and reading can be done simultaneously without changing the position.

Major parts of a Prismatic Compass are:

- Magnetic needle.

- Graduated ring.

- Adjustable mirror.

- Sliding arrangement for mirror.

- Object vane.

- Eye vane.

- Metal box.

- Glass cover.

- Horse hair.

Prismatic compass (Geology Superstore).

Adjustments of Prismatic Compass

Two types of adjustments:

- Temporary adjustment.

- Permanent adjustment.

Temporary Adjustments

- Centering: It is the process of fixing the compass exactly over the station. Centering is usually done by adjusting the tripod legs. Also a plumb-bob is used to judge the accurate centering of instruments over the station.

- Leveling: The instrument has to be leveled if it is used as in hand or mounted over a tripod. If it is used as in hand, the graduated disc should swing freely and appears to be completely level in reference to the top edge of the case. If the tripod is used, they usually have a ball and socket arrangement for leveling purpose.

- Focusing the prism: Prism can be slide up or down for focusing to make the readings clear and readable.

  Permanent adjustments are same as in the Surveyor's compass.

Surveyor's Compass

Surveyor's compass consists of a circular brass box containing a magnetic needle which swings freely over a brass circle which is divided into 360 degrees. The horizontal angle is measured using a pair of sights located on north – south axis of the compass. They are usually mounted over a tripod and leveled using a ball and socket mechanism.

Surveyor's compass (National Museum of American History).

They also have two types of adjustments, temporary and permanent. Temporary adjustments are same as described in prismatic compass.

Permanent Adjustments

They are done only in the circumstances where the internal parts of the prism is disturbed or damaged. They are:

- Adjustments in levels,
- Adjustment of pivot point,
- Adjustment of sight vanes,
- Adjustment of needle.

Measurement of Angles and Computation of Area

The observations of a plot using compass surveying will be:

Here the bearings are observed in Whole Circle Bearing (W.C.B) system.

Line	Fore bearing	Distance measured (m)
AB	40° 0'	10.8
BC	110°0'	8.2
CA	280°0'	13.1

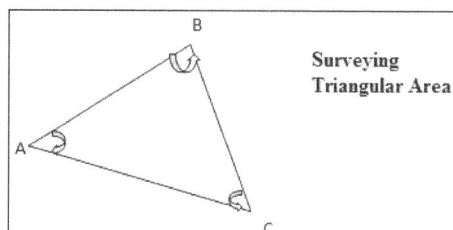

Surveying Triangular Area

Included angle = Bearing of previous line – Bearing of next line

$= (280°-180°) -40° = 60°$

$Sum = 180°$

Check,

$(2n-4) 90° = (6-4) 90° = 180°.$

Where, n = number of sides of the traverse.

Area Computation

$$Area = \frac{ab \sin c}{2} = \frac{10.8 + 13.1 + \sin 60°}{2} = 61.2 \, m^2$$

Advantages and Disadvantages of Compass Surveying

Advantages

- They are portable and light weight.
- They have fewer settings to fix it on a station.
- The error in direction produced in a single survey line does not affect other lines.
- It is suitable to retrace old surveys.

Disadvantages

- It is less precise compared to other advanced methods of surveying.
- It is easily subjected to various errors such as errors adjoining to magnetic meridian, local attraction etc.
- Imperfect sighting of the ranging rods and inaccurate leveling also causes error.

Errors in Compass Survey

Errors can be arising due to various reasons during the process of surveying, they are classified as:

- Instrumental errors.
- Personal errors.
- Natural errors.

Instrumental Errors

As the name suggests they are arise due to the wrong adjustments of the instruments. Some other reasons are:

- If the plane of sight not being vertical, it causes error in sighting and reading.

- If the magnetic needle is not perfectly straight or if it is sluggish, readings may not be accurate.

Personal Errors

They arise mainly due to the carelessness of the surveyor. They are:

- Inaccurate leveling,
- Inaccurate reading,
- Inaccurate centering.

Natural Errors

Natural errors are occurring due to the various natural causes which affect the working of compass. It has nothing to do with the surveyor and to minimize them, some corrections in calculations applied. They are:

- Local attraction,
- Proximity to the magnetic storms,
- Declination.

Remote Sensing

Remote sensing is the acquisition of information about an object or phenomenon without making physical contact with the object and thus in contrast to on-site observation, especially the Earth. Remote sensing is used in numerous fields, including geography, land surveying and most Earth science disciplines (for example, hydrology, ecology, meteorology, oceanography, glaciology, geology); it also has military, intelligence, commercial, economic, planning, and humanitarian applications.

In current usage, the term "remote sensing" generally refers to the use of satellite- or aircraft-based sensor technologies to detect and classify objects on Earth, including on the surface and in the atmosphere and oceans, based on propagated signals (e.g. electromagnetic radiation). It may be split into "active" remote sensing (such as when a signal is emitted by a satellite or aircraft and its reflection by the object is detected by the sensor) and "passive" remote sensing (such as when the reflection of sunlight is detected by the sensor).

Passive sensors gather radiation that is emitted or reflected by the object or surrounding areas. Reflected sunlight is the most common source of radiation measured by passive sensors. Examples of passive remote sensors include film photography, infrared, charge-coupled devices, and radiometers. Active collection, on the other hand, emits energy in order to scan objects and areas whereupon a sensor then detects and measures the radiation that is reflected or backscattered from the target. RADAR and LIDAR are examples of active remote sensing where the time delay between emission and return is measured, establishing the location, speed and direction of an object.

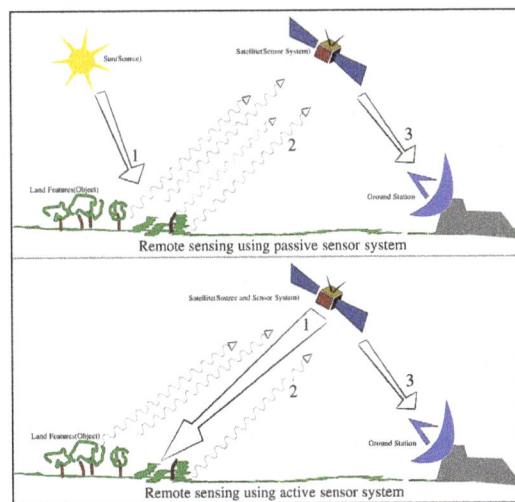

Illustration of remote sensing.

Remote sensing makes it possible to collect data of dangerous or inaccessible areas. Remote sensing applications include monitoring deforestation in areas such as the Amazon Basin, glacial features in Arctic and Antarctic regions, and depth sounding of coastal and ocean depths. Military collection during the Cold War made use of stand-off collection of data about dangerous border areas. Remote sensing also replaces costly and slow data collection on the ground, ensuring in the process that areas or objects are not disturbed.

Orbital platforms collect and transmit data from different parts of the electromagnetic spectrum, which in conjunction with larger scale aerial or ground-based sensing and analysis, provides researchers with enough information to monitor trends such as El Niño and other natural long and short term phenomena. Other uses include different areas of the earth sciences such as natural resource management, agricultural fields such as land usage and conservation, and national security and overhead, ground-based and stand-off collection on border areas.

Types of Data Acquisition Techniques

The basis for multispectral collection and analysis is that of examined areas or objects that reflect or emit radiation that stand out from surrounding areas.

Applications of Remote Sensing

- Conventional radar is mostly associated with aerial traffic control, early warning, and certain large scale meteorological data. Doppler radar is used by local law enforcements' monitoring of speed limits and in enhanced meteorological collection such as wind speed and direction within weather systems in addition to precipitation location and intensity. Other types of active collection includes plasmas in the ionosphere. Interferometric synthetic aperture radar is used to produce precise digital elevation models of large scale terrain.

- Laser and radar altimeters on satellites have provided a wide range of data. By measuring the bulges of water caused by gravity, they map features on the seafloor to a resolution of a mile or so. By measuring the height and wavelength of ocean waves, the altimeters measure wind speeds and direction, and surface ocean currents and directions.

- Ultrasound (acoustic) and radar tide gauges measure sea level, tides and wave direction in coastal and offshore tide gauges.

- Light detection and ranging (LIDAR) is well known in examples of weapon ranging, laser illuminated homing of projectiles. LIDAR is used to detect and measure the concentration of various chemicals in the atmosphere, while airborne LIDAR can be used to measure heights of objects and features on the ground more accurately than with radar technology. Vegetation remote sensing is a principal application of LIDAR.

- Radiometers and photometers are the most common instrument in use, collecting reflected and emitted radiation in a wide range of frequencies. The most common are visible and infrared sensors, followed by microwave, gamma ray and rarely, ultraviolet. They may also be used to detect the emission spectra of various chemicals, providing data on chemical concentrations in the atmosphere.

- Spectropolarimetric Imaging has been reported to be useful for target tracking purposes by researchers at the U.S. Army Research Laboratory. They determined that manmade items possess polarimetric signatures that are not found in natural objects. These conclusions were drawn from the imaging of military trucks, like the Humvee, and trailers with their acousto-optic tunable filter dual hyperspectral and spectropolarimetric VNIR Spectropolarimetric Imager.

- Stereographic pairs of aerial photographs have often been used to make topographic maps by imagery and terrain analysts in trafficability and highway departments for potential routes, in addition to modelling terrestrial habitat features.

- Simultaneous multi-spectral platforms such as Landsat have been in use since the 1970s. These thematic mappers take images in multiple wavelengths of electro-magnetic radiation (multi-spectral) and are usually found on Earth observation satellites, including (for example) the Landsat program or the IKONOS satellite. Maps of land cover and land use from thematic mapping can be used to prospect for minerals, detect or monitor land usage, detect invasive vegetation, deforestation, and examine the health of indigenous plants and crops, including entire farming regions or forests. Prominent scientists using remote sensing for this purpose include Janet Franklin and Ruth DeFries. Landsat images are used by regulatory agencies such as KYDOW to indicate water quality parameters including Secchi depth, chlorophyll a density and total phosphorus content. Weather satellites are used in meteorology and climatology.

- Hyperspectral imaging produces an image where each pixel has full spectral information with imaging narrow spectral bands over a contiguous spectral range. Hyperspectral imagers are used in various applications including mineralogy, biology, defence, and environmental measurements.

- Within the scope of the combat against desertification, remote sensing allows researchers to follow up and monitor risk areas in the long term, to determine desertification factors, to support decision-makers in defining relevant measures of environmental management, and to assess their impacts.

Geodetic

- Geodetic remote sensing can be gravimetric or geometric. Overhead gravity data collection was first used in aerial submarine detection. This data revealed minute perturbations in the Earth's gravitational field that may be used to determine changes in the mass distribution of the Earth, which in turn may be used for geophysical studies, as in GRACE. Geometric remote sensing includes position and deformation imaging using InSAR, LIDAR, etc.

Acoustic and Near-acoustic

- Sonar: Passive sonar, listening for the sound made by another object (a vessel, a whale etc.); *active sonar*, emitting pulses of sounds and listening for echoes, used for detecting, ranging and measurements of underwater objects and terrain.

- Seismograms taken at different locations can locate and measure earthquakes (after they occur) by comparing the relative intensity and precise timings.

- Ultrasound: Ultrasound sensors, that emit high frequency pulses and listening for echoes, used for detecting water waves and water level, as in tide gauges or for towing tanks.

To coordinate a series of large-scale observations, most sensing systems depend on the following: platform location and the orientation of the sensor. High-end instruments now often use positional information from satellite navigation systems. The rotation and orientation is often provided within a degree or two with electronic compasses. Compasses can measure not just azimuth (i. e. degrees to magnetic north), but also altitude (degrees above the horizon), since the magnetic field curves into the Earth at different angles at different latitudes. More exact orientations require gyroscopic-aided orientation, periodically realigned by different methods including navigation from stars or known benchmarks.

Data Characteristics

The quality of remote sensing data consists of its spatial, spectral, radiometric and temporal resolutions.

Spatial Resolution

The size of a pixel that is recorded in a raster image – typically pixels may correspond to square areas ranging in size length from 1 to 1,000 metres (3.3 to 3,280.8 ft).

Spectral Resolution

The wavelength of the different frequency bands recorded – usually, this is related to the number of frequency bands recorded by the platform. Current Landsat collection is that of seven bands, including several in the infrared spectrum, ranging from a spectral resolution of 0.7 to 2.1 μm. The Hyperion sensor on Earth Observing-1 resolves 220 bands from 0.4 to 2.5 μm, with a spectral resolution of 0.10 to 0.11 μm per band.

Radiometric Resolution

The number of different intensities of radiation the sensor is able to distinguish. Typically, this ranges from 8 to 14 bits, corresponding to 256 levels of the gray scale and up to 16,384 intensities or "shades" of colour, in each band. It also depends on the instrument noise.

Temporal Resolution

The frequency of flyovers by the satellite or plane, and is only relevant in time-series studies or those requiring an averaged or mosaic image as in deforesting monitoring. This was first used by the intelligence community where repeated coverage revealed changes in infrastructure, the deployment of units or the modification/introduction of equipment. Cloud cover over a given area or object makes it necessary to repeat the collection of said location.

Data Processing

In order to create sensor-based maps, most remote sensing systems expect to extrapolate sensor data in relation to a reference point including distances between known points on the ground. This depends on the type of sensor used. For example, in conventional photographs, distances are accurate in the center of the image, with the distortion of measurements increasing the farther you get from the center. Another factor is that of the platen against which the film is pressed can cause severe errors when photographs are used to measure ground distances. The step in which this problem is resolved is called georeferencing, and involves computer-aided matching of points in the image (typically 30 or more points per image) which is extrapolated with the use of an established benchmark, "warping" the image to produce accurate spatial data. As of the early 1990s, most satellite images are sold fully georeferenced.

In addition, images may need to be radiometrically and atmospherically corrected.

Radiometric Correction

Allows avoidance of radiometric errors and distortions. The illumination of objects on the Earth surface is uneven because of different properties of the relief. This factor is taken into account in the method of radiometric distortion correction. Radiometric correction gives a scale to the pixel values, e. g. the monochromatic scale of 0 to 255 will be converted to actual radiance values.

Topographic Correction (also called Terrain Correction)

In rugged mountains, as a result of terrain, the effective illumination of pixels varies considerably. In a remote sensing image, the pixel on the shady slope receives weak illumination and has a low radiance value, in contrast, the pixel on the sunny slope receives strong illumination and has a high radiance value. For the same object, the pixel radiance value on the shady slope will be different from that on the sunny slope. Additionally, different objects may have similar radiance values. These ambiguities seriously affected remote sensing image information extraction accuracy in mountainous areas. It became the main obstacle to further application of remote sensing images. The purpose of topographic correction is to eliminate this effect, recovering the true reflectivity or radiance of objects in horizontal conditions. It is the premise of quantitative remote sensing application.

Atmospheric Correction

Elimination of atmospheric haze by rescaling each frequency band so that its minimum value (usually realised in water bodies) corresponds to a pixel value of 0. The digitizing of data also makes it possible to manipulate the data by changing gray-scale values.

Interpretation is the critical process of making sense of the data. The first application was that of aerial photographic collection which used the following process; spatial measurement through the use of a light table in both conventional single or stereographic coverage, added skills such as the use of photogrammetry, the use of photomosaics, repeat coverage, making use of objects' known dimensions in order to detect modifications. Image Analysis is the recently developed automated computer-aided application which is in increasing use.

Object-Based Image Analysis (OBIA) is a sub-discipline of GIScience devoted to partitioning remote sensing (RS) imagery into meaningful image-objects, and assessing their characteristics through spatial, spectral and temporal scale.

Old data from remote sensing is often valuable because it may provide the only long-term data for a large extent of geography. At the same time, the data is often complex to interpret, and bulky to store. Modern systems tend to store the data digitally, often with lossless compression. The difficulty with this approach is that the data is fragile, the format may be archaic, and the data may be easy to falsify. One of the best systems for archiving data series is as computer-generated machine-readable ultrafiche, usually in typefonts such as OCR-B, or as digitized half-tone images. Ultrafiches survive well in standard libraries, with lifetimes of several centuries. They can be created, copied, filed and retrieved by automated systems. They are about as compact as archival magnetic media, and yet can be read by human beings with minimal, standardized equipment.

Generally speaking, remote sensing works on the principle of the *inverse problem*: while the object or phenomenon of interest (the state) may not be directly measured, there exists some other variable that can be detected and measured (the observation) which may be related to the object of interest through a calculation. The common analogy given to describe this is trying to determine the type of animal from its footprints. For example, while it is impossible to directly measure temperatures in the upper atmosphere, it is possible to measure the spectral emissions from a known chemical species (such as carbon dioxide) in that region. The frequency of the emissions may then be related via thermodynamics to the temperature in that region.

Data Processing Levels

To facilitate the discussion of data processing in practice, several processing "levels" were first defined in 1986 by NASA as part of its Earth Observing System and steadily adopted since then, both internally at NASA and elsewhere; these definitions are:

Level	Description
0	Reconstructed, unprocessed instrument and payload data at full resolution, with any and all communications artifacts (e. g., synchronization frames, communications headers, duplicate data) removed.
1a	Reconstructed, unprocessed instrument data at full resolution, time-referenced, and annotated with ancillary information, including radiometric and geometric calibration coefficients and georeferencing parameters (e. g., platform ephemeris) computed and appended but not applied to the Level 0 data (or if applied, in a manner that level 0 is fully recoverable from level 1a data).

1b	Level 1a data that have been processed to sensor units (e. g., radar backscatter cross section, brightness temperature, etc.); not all instruments have Level 1b data; level 0 data is not recoverable from level 1b data.
2	Derived geophysical variables (e. g., ocean wave height, soil moisture, ice concentration) at the same resolution and location as Level 1 source data.
3	Variables mapped on uniform spacetime grid scales, usually with some completeness and consistency (e. g., missing points interpolated, complete regions mosaicked together from multiple orbits, etc.).
4	Model output or results from analyses of lower level data (i. e., variables that were not measured by the instruments but instead are derived from these measurements).

A Level 1 data record is the most fundamental (i. e., highest reversible level) data record that has significant scientific utility, and is the foundation upon which all subsequent data sets are produced. Level 2 is the first level that is directly usable for most scientific applications; its value is much greater than the lower levels. Level 2 data sets tend to be less voluminous than Level 1 data because they have been reduced temporally, spatially, or spectrally. Level 3 data sets are generally smaller than lower level data sets and thus can be dealt with without incurring a great deal of data handling overhead. These data tend to be generally more useful for many applications. The regular spatial and temporal organization of Level 3 datasets makes it feasible to readily combine data from different sources.

While these processing levels are particularly suitable for typical satellite data processing pipelines, other data level vocabularies have been defined and may be appropriate for more heterogeneous workflows.

Software

Remote sensing data are processed and analyzed with computer software, known as a remote sensing application. A large number of proprietary and open source applications exist to process remote sensing data. Remote sensing software packages include:

- ERDAS IMAGINE from Hexagon Geospatial (Separated from Intergraph SG&I),

- PCI Geomatica,

- TNTmips from MicroImages,

- IDRISI from Clark Labs,

- eCognition from Trimble,

- and RemoteView made by Overwatch Textron Systems,

- Dragon/ips is one of the oldest remote sensing packages still available, and is in some cases free.

Open source remote sensing software includes:

- Opticks (software),

- Orfeo toolbox,

- Sentinel Application Platform (SNAP) from the European Space Agency (ESA),

- Others mixing remote sensing and GIS capabilities are: GRASS GIS, ILWIS, QGIS, and TerraLook.

According to an NOAA Sponsored Research by Global Marketing Insights, Inc. the most used applications among Asian academic groups involved in remote sensing are as follows: ERDAS 36% (ERDAS IMAGINE 25% & ERMapper 11%); ESRI 30%; ITT Visual Information Solutions ENVI 17%; MapInfo 17%.

Among Western Academic respondents as follows: ESRI 39%, ERDAS IMAGINE 27%, MapInfo 9%, and AutoDesk 7%.

In education, those that want to go beyond simply looking at satellite images print-outs either use general remote sensing software (e.g. QGIS), Google Earth, StoryMaps or a software/web-app developed specifically for education (e.g. desktop: LeoWorks, online: BLIF).

Satellites

First satellite UV/VIS observations simply showed pictures of the Earth's surface and atmosphere. Such satellite images are still used, for instance as input for numerical weather forecast. The first spectroscopic UV/VIS observations started in 1970 on board of the US research satellite Nimbus 4. These measurements (backscatter ultraviolet, BUV, later also called Solar BUV, SBUV) operated in nadir geometry, i.e., they measured the solar light reflected from the ground or scattered from the atmosphere. Like for the Dobson instruments also the BUV/SBUV instruments measure the intensity in different narrow spectral intervals. The intention of these BUV/SBUV observations was to determine information on the atmospheric O_3 profile, since the penetration depth into the atmosphere strongly depends on wavelength. For example, the light at the shortest wavelengths has only 'seen' the highest parts of the O_3 layer whereas the longest wavelengths have seen the total column. While in principle the BUV/SBUV measurements worked well, they suffered from instrumental instabilities.

The big breakthrough in UV/VIS satellite remote sensing of the atmosphere took place in 1979 with the launch the Total Ozone Mapping Spectrometer (TOMS) on Nimbus 7. TOMS is similar to the BUV/SBUV instrument but measures light at longer wavelengths. Thus it is only sensitive to the total O_3 column (instead of the O_3 profile). However, compared to the BUV/SBUV instruments the TOMS instruments were much more stable. The TOMS instrument on board of Nimbus 7 yielded the so far longest global data set on O_3 (1979–1992). This period in particular includes the discovery of the ozone hole. Several further TOMS instruments have been launched on other satellites. Like the Dobson instruments on the ground they yield very accurate O_3 total column densities using a relatively simple method. Besides events of very strong atmospheric SO_2 absorption and aerosols they are, however, only sensitive to O_3.

Since April 1995 the first DOAS instrument is operating from space. The Global Ozone Monitoring Experiment (GOME) was launched on the European research satellite ERS-2. Like SBUV and TOMS also GOME is a nadir viewing instrument; unlike its predecessor instruments it covers a large spectral range (240 – 790 nm) at a total of 4096 wavelengths arranged in four 'channels' with a spectral resolution between 0.2 and 0.4 nm. Its normal ground pixel size in 320 × 40 km². Global coverage is achieved after three days. For O_3 profile measurements the intensities at short wavelengths are observed (BUV/SBUV instruments). For the determination of the total atmospheric O_3 column the intensities at larger wavelengths are used (TOMS instruments). In contrast to the limited spectral information of BUV/SBUV and TOMS instruments, GOME spectra yield a surplus of spectral information. By applying the DOAS method to these measurements it is thus possible to retrieve a large variety of atmospheric trace gases, the majority of which are very weak absorbers (O_3, NO_2, BrO, OClO, HCHO, H_2O, O_2, O_4, SO_2). In addition other quantities like aerosol absorptions, the ground albedo or indices characterising the solar cycle can be analysed. Because of the high sensitivity of GOME it is in particular possible to measure various tropospheric trace gases (NO_2, BrO, HCHO, H2O, SO_2). A further important advantage is that the GOME spectra can be analysed with respect to a spectrum of direct sun light, which contains no atmospheric absorptions. Therefore, in contrast to ground based DOAS measurements the DOAS analysis of GOME spectra yields total atmospheric column densities rather than the difference between two atmospheric spectra.

In March 2002 a second DOAS satellite instrument, the SCanning Imaging Absorption SpectroMeter for Atmospheric ChartographY (SCIAMACHY) was launched on board of the European research satellite Envisat. In addition to GOME it measures over a wider wavelength range (240 nm – 2380) including also the absorption of several greenhouse gases (CO_2, CH_4, N_2O) and CO in the IR. It also operated in additional viewing modes (nadir, limb, occultation), which allows to derive stratospheric trace gas profiles. Another advantage is that the ground pixel size for the nadir viewing mode was significantly reduced to 30 x 60 km² (in a special mode even to 15 x 30 km²). Especially for the observation of tropospheric trace gases this is very important because of the strong spatial gradients occurring for such species. The first tropospheric results of SCIAMACHY showed that it was now possible to identify pollution plumes of individual cities or other big sources.

Photogrammetry

Photogrammetry is the art, science and technology of obtaining reliable information about physical objects and the environment through the process of recording, measuring and interpreting photographic images and patterns of electromagnetic radiant imagery and other phenomena. A simplified definition could be the extraction of three-dimensional measurements from two-dimensional data (i.e. images); for example, the distance between two points that lie on a plane parallel to the photographic image plane can be determined by measuring their distance on the image, if the scale of the image is known. Close-range photogrammetry refers to the collection of photography from a lesser distance than traditional aerial (or orbital) photogrammetry. Photogrammetry is as old as modern photography, dating to the mid-19th century.

Photogrammetric analysis may be applied to one photograph, or may use high-speed photography and remote sensing to detect, measure and record complex 2D and 3D motion fields by feeding measurements and imagery analysis into computational models in an attempt to successively estimate, with increasing accuracy, the actual, 3D relative motions.

From its beginning with the stereoplotters used to plot contour lines on topographic maps, it now has a very wide range of uses such as sonar, radar, and lidar.

Methods

Photogrammetry uses methods from many disciplines, including optics and projective geometry. Digital image capturing and photogrammetric processing includes several well defined stages, which allow the generation of 2D or 3D digital models of the object as an end product. The data model on the right shows what type of information can go into and come out of photogrammetric methods.

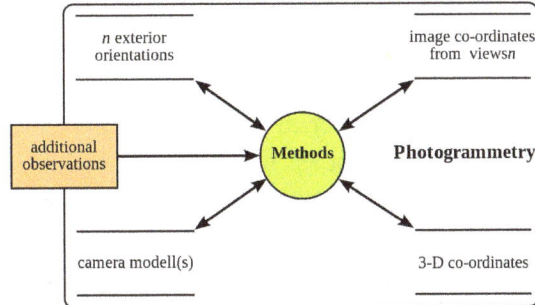

A data model of photogrammetry.

The 3D co-ordinates define the locations of object points in the 3D space. The image co-ordinates define the locations of the object points' images on the film or an electronic imaging device. The exterior orientation of a camera defines its location in space and its view direction. The inner orientation defines the geometric parameters of the imaging process. This is primarily the focal length of the lens, but can also include the description of lens distortions. Further additional observations play an important role: With scale bars, basically a known distance of two points in space, or known fix points, the connection to the basic measuring units is created.

Each of the four main variables can be an *input* or an *output* of a photogrammetric method.

Algorithms for photogrammetry typically attempt to minimize the sum of the squares of errors over the coordinates and relative displacements of the reference points. This minimization is known as bundle adjustment and is often performed using the Levenberg–Marquardt algorithm.

Stereophotogrammetry

A special case, called stereophotogrammetry, involves estimating the three-dimensional coordinates of points on an object employing measurements made in two or more photographic images taken from different positions. Common points are identified on each image. A line of sight (or ray) can be constructed from the camera location to the point on the object. It is the intersection of these rays (triangulation) that determines the three-dimensional location of the point. More sophisticated algorithms can exploit other information about the scene that is known *a priori*, for example symmetries, in some cases allowing reconstructions of 3D coordinates from only one camera position. Stereophotogrammetry is emerging as a robust non-contacting measurement technique to determine dynamic characteristics and mode shapes of non-rotating and rotating structures.

Integration

Photogrammetric data with a dense range data in which scanners complement each other. Photogrammetry is more accurate in the x and y direction while range data are generally more accurate in the z direction. This range data can be supplied by techniques like LIDAR, laser scanners (using time of flight, triangulation or interferometry), white-light digitizers and any other technique that scans an area and returns x, y, z coordinates for multiple discrete points (commonly called "point clouds"). Photos can clearly define the edges of buildings when the point cloud footprint can not. It is beneficial to incorporate the advantages of both systems and integrate them to create a better product.

A 3D visualization can be created by georeferencing the aerial photos and LIDAR data in the same reference frame, orthorectifying the aerial photos, and then draping the orthorectified images on top of the LIDAR grid. It is also possible to create digital terrain models and thus 3D visualisations using pairs (or multiples) of aerial photographs or satellite (e.g. SPOT satellite imagery). Techniques such as adaptive least squares stereo matching are then used to produce a dense array of correspondences which are transformed through a camera model to produce a dense array of x, y, z data which can be used to produce digital terrain model and orthoimage products. Systems which use these techniques, e.g. the ITG system, were developed in the 1980s and 1990s but have since been supplanted by LIDAR and radar-based approaches, although these techniques may still be useful in deriving elevation models from old aerial photographs or satellite images.

Applications

Photogrammetry is used in fields such as topographic mapping, architecture, engineering, manufacturing, quality control, police investigation, cultural heritage, and geology. Archaeologists use it to quickly produce plans of large or complex sites, and meteorologists use it to determine the wind speed of tornados when objective weather data cannot be obtained.

Photograph of person using controller to explore a 3D Photogrammetry experience, Future Cities by DERIVE, recreating Tokyo.

It is also used to combine live action with computer-generated imagery in movies post-production; The Matrix is a good example of the use of photogrammetry in film. Photogrammetry was used extensively to create photorealistic environmental assets for video games including The Vanishing of Ethan Carter as well as EA DICE's Star Wars Battlefront. The main character of the game Hellblade: Senua's Sacrifice was derived from photogrammetric motion-capture models taken of actress Melina Juergens.

Photogrammetry is also commonly employed in collision engineering, especially with automobiles. When litigation for accidents occurs and engineers need to determine the exact deformation present in the vehicle, it is common for several years to have passed and the only evidence that remains is accident scene photographs taken by the police. Photogrammetry is used to determine how much the car in question was deformed, which relates to the amount of energy required to produce that deformation. The energy can then be used to determine important information about the crash (such as the velocity at time of impact).

Mapping

Photomapping is the process of making a map with "cartographic enhancements" that have been drawn from a photomosaic that is "a composite photographic image of the ground" or more precisely as a controlled photomosaic where "individual photographs are rectified for tilt and brought to a common scale (at least at certain control points)."

Rectification of imagery is generally achieved by "fitting the projected images of each photograph to a set of four control points whose positions have been derived from an existing map or from ground measurements. When these rectified, scaled photographs are positioned on a grid of control points, a good correspondence can be achieved between them through skillful trimming and fitting and the use of the areas around the principal point where the relief displacements (which cannot be removed) are at a minimum."

"It is quite reasonable to conclude that some form of photomap will become the standard general map of the future." go on to suggest that, "photomapping would appear to be the only way to take reasonable advantage" of future data sources like high altitude aircraft and satellite imagery. The highest resolution aerial photomaps on GoogleEarth are approximately 2.5 cm (0.98 in) spatial resolution images. The highest resolution photomap of ortho images was made in Hungary in 2012 with a 0.5 cm (0.20 in) spatial resolution.

Archaeology

Using a pentop computer to photomap an archaeological excavation in the field.

Demonstrating the link between orthophotomapping and archaeology, historic airphotos photos were used to aid in developing a reconstruction of the Ventura mission that guided excavations of the structure's walls.

Overhead photography has been widely applied for mapping surface remains and excavation exposures at archaeological sites. Suggested platforms for capturing these photographs has included: War Balloons from World War I; rubber meteorological balloons; kites; wooden platforms, metal frameworks, constructed over an excavation exposure; ladders both alone and held together with poles or planks; three legged ladders; single and multi-section poles; bipods; tripods; tetrapods aerial bucket trucks ("cherry pickers"), and light weight individuals dangling from the limb of a nearby tree.

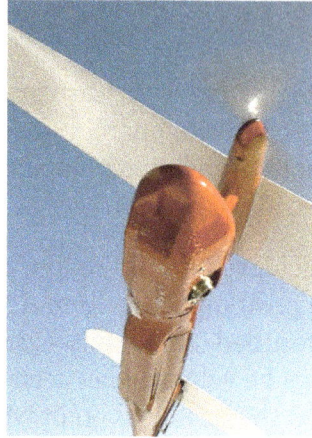

Pteryx UAV, a civilian UAV for aerial photography and
photomapping with roll-stabilised camera head.

Hand held near nadir overhead digital photographs have been used with geographic information systems (GIS) to record excavation exposures.

Photogrammetry is increasingly being used in maritime archaeology because of the relative ease of mapping sites compared to traditional methods, allowing the creation of 3D maps which can be rendered in virtual reality.

3D Modeling

A somewhat similar application is the scanning of objects to automatically make 3D models of them. The produced model often still contains gaps, so additional cleanup with software like MeshLab, netfabb or MeshMixer is often still necessary.

Satellite Imagery

Satellite imagery (also Earth observation imagery or spaceborne photography) are images of Earth or other planets collected by imaging satellites operated by governments and businesses around the world. Satellite imaging companies sell images by licensing them to governments and businesses such as Apple Maps and Google Maps.

The first images from space were taken on sub-orbital flights. The U.S-launched V-2 flight on October 24, 1946 took one image every 1.5 seconds. With an apogee of 65 miles (105 km), these photos were from five times higher than the previous record, the 13.7 miles (22 km) by the Explorer II balloon mission in 1935. The first satellite (orbital) photographs of Earth were made on August 14, 1959 by the U.S. Explorer 6. The first satellite photographs of the Moon might have been made on October 6, 1959 by the Soviet satellite Luna 3, on a mission to photograph the far side of the Moon. The Blue Marble photograph was taken from space in 1972, and has become very popular in the media and among the public. Also in 1972 the United States started the Landsat program, the largest program for acquisition of imagery of Earth from space. Landsat Data Continuity Mission, the most recent Landsat satellite, was launched on 11 February 2013. In 1977, the first real time satellite imagery was acquired by the United States's KH-11 satellite system.

The satellite images were made from pixels.

The first crude image taken by the satellite Explorer 6 shows a sunlit area of the Central Pacific Ocean and its cloud cover. The photo was taken when the satellite was about 17,000 mi (27,000 km) above the surface of the earth on August 14, 1959. At the time, the satellite was crossing Mexico.

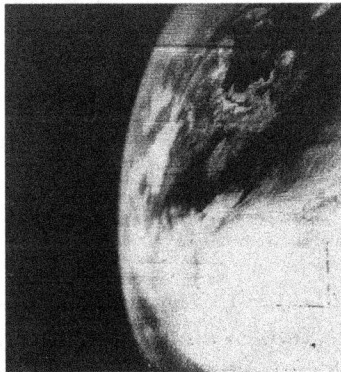

The first television image of Earth from space
transmitted by the TIROS-1 weather satellite.

All satellite images produced by NASA are published by NASA Earth Observatory and are freely available to the public. Several other countries have satellite imaging programs, and a collaborative European effort launched the ERS and Envisat satellites carrying various sensors. There are also private companies that provide commercial satellite imagery. In the early 21st century satellite imagery became widely available when affordable, easy to use software with access to satellite imagery databases was offered by several companies and organizations.

Uses

Satellite images have many applications in meteorology, oceanography, fishing, agriculture, biodiversity conservation, forestry, landscape, geology, cartography, regional planning, education, intelligence and warfare. Images can be in visible colors and in other spectra. There are also elevation maps, usually made by radar images. Interpretation and analysis of satellite imagery is conducted using specialized remote sensing software.

Satellite photography can be used to produce composite images of an entire hemisphere.

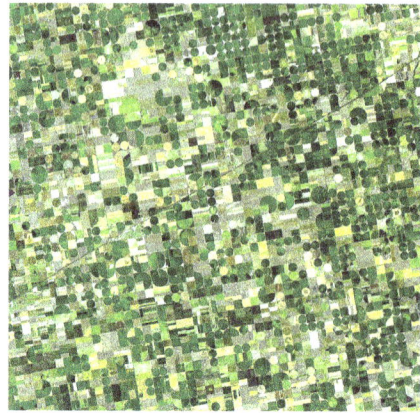

Or to map a small area of the Earth, such as this photo of the countryside of Haskell County, Kansas, United States.

Data Characteristics

There are four types of resolution when discussing satellite imagery in remote sensing: spatial, spectral, temporal, and radiometric. Campbell defines these as follows:

- Spatial resolution is defined as the pixel size of an image representing the size of the surface area (i.e. M^2) being measured on the ground, determined by the sensors' instantaneous field of view (ifov).

- Spectral resolution is defined by the wavelength interval size (discrete segment of the electromagnetic spectrum) and number of intervals that the sensor is measuring.

- Temporal resolution is defined by the amount of time (e.G. Days) that passes between imagery collection periods for a given surface location.

- Radiometric resolution is defined as the ability of an imaging system to record many levels of brightness (contrast for example) and to the effective bit-depth of the sensor (number of grayscale levels) and is typically expressed as 8-bit (0–255), 11-bit (0–2047), 12-bit (0–4095) or 16-bit (0–65,535).

- Geometric resolution refers to the satellite sensor's ability to effectively image a portion of the Earth's surface in a single pixel and is typically expressed in terms of Ground sample distance, or GSD. GSD is a term containing the overall optical and systemic noise sources and is useful for comparing how well one sensor can "see" an object on the ground within a single pixel. For example, the GSD of Landsat is ≈30m, which means the smallest unit that maps to a single pixel within an image is ≈30m x 30m. The latest commercial satellite (GeoEye 1) has a GSD of 0.41 m. This compares to a 0.3 m resolution obtained by some early military film based Reconnaissance satellite such as Corona.

The resolution of satellite images varies depending on the instrument used and the altitude of the satellite's orbit. For example, the Landsat archive offers repeated imagery at 30 meter resolution for the planet, but most of it has not been processed from the raw data. Landsat 7 has an average return period of 16 days. For many smaller areas, images with resolution as high as 41 cm can be available.

Satellite imagery is sometimes supplemented with aerial photography, which has higher resolution, but is more expensive per square meter. Satellite imagery can be combined with vector or raster data in a GIS provided that the imagery has been spatially rectified so that it will properly align with other data sets.

Imaging Satellites

Public Domain

Satellite imaging of the Earth surface is of sufficient public utility that many countries maintain satellite imaging programs. The United States has led the way in making these data freely available for scientific use. Some of the more popular programs are listed below, recently followed by the European Union's Sentinel constellation.

Landsat

Landsat is the oldest continuous Earth observing satellite imaging program. Optical Landsat imagery has been collected at 30 m resolution since the early 1980s. Beginning with Landsat 5, thermal infrared imagery was also collected (at coarser spatial resolution than the optical data). The Landsat 7 and Landsat 8 satellites are currently in orbit. Landsat 9 is planned.

Modis

MODIS has collected near-daily satellite imagery of the earth in 36 spectral bands since 2000. MODIS is on board the NASA Terra and Aqua satellites.

Sentinel

The ESA is currently developing the Sentinel constellation of satellites. Currently, 7 missions are planned, each for a different application. Sentinel-1 (SAR imaging), Sentinel-2 (decameter optical imaging for land surfaces), and Sentinel-3 (hectometer optical and thermal imaging for land and water) have already been launched.

Aster

The Advanced Spaceborne Thermal Emission and Reflection Radiometer (ASTER) is an imaging instrument onboard Terra, the flagship satellite of NASA's Earth Observing System (EOS) launched in December 1999. ASTER is a cooperative effort between NASA, Japan's Ministry of Economy, Trade and Industry (METI), and Japan Space Systems (J-spacesystems). ASTER data is used to create detailed maps of land surface temperature, reflectance, and elevation. The coordinated system of EOS satellites, including Terra, is a major component of NASA's Science Mission Directorate and the Earth Science Division. The goal of NASA Earth Science is to develop a scientific understanding of the Earth as an integrated system, its response to change, and to better predict variability and trends in climate, weather, and natural hazards.

- Land surface climatology: Investigation of land surface parameters, surface temperature, etc., to understand land-surface interaction and energy and moisture fluxes.

- Vegetation and ecosystem dynamics: Investigations of vegetation and soil distribution and their changes to estimate biological productivity, understand land-atmosphere interactions, and detect ecosystem change.

- Volcano monitoring: Monitoring of eruptions and precursor events, such as gas emissions, eruption plumes, development of lava lakes, eruptive history and eruptive potential.

- Hazard monitoring: Observation of the extent and effects of wildfires, flooding, coastal erosion, earthquake damage, and tsunami damage.

- Hydrology: Understanding global energy and hydrologic processes and their relationship to global change; included is evapotranspiration from plants.

- Geology and soils: The detailed composition and geomorphologic mapping of surface soils and bedrocks to study land surface processes and earth's history.

- Land surface and land cover change: Monitoring desertification, deforestation, and urbanization; providing data for conservation managers to monitor protected areas, national parks, and wilderness areas.

Meteosat

Model of a first generation Meteosat geostationary satellite.

The Meteosat-2 geostationary weather satellite began operationally to supply imagery data on 16 August 1981. Eumetsat has operated the Meteosats since 1987.

- The Meteosat visible and infrared imager (MVIRI), three-channel imager: visible, infrared and water vapour; It operates on the first generation Meteosat, Meteosat-7 being still active.

- The 12-channel Spinning Enhanced Visible and Infrared Imager (SEVIRI) includes similar channels to those used by MVIRI, providing continuity in climate data over three decades; Meteosat Second Generation (MSG).

- The Flexible Combined Imager (FCI) on Meteosat Third Generation (MTG) will also include similar channels, meaning that all three generations will have provided over 60 years of climate data.

Private Domain

Several satellites are built and maintained by private companies. These include:

GeoEye

GeoEye's GeoEye-1 satellite was launched on September 6, 2008. The GeoEye-1 satellite has the high resolution imaging system and is able to collect images with a ground resolution of 0.41 meters (16 inches) in the panchromatic or black and white mode. It collects multispectral or color imagery at 1.65-meter resolution or about 64 inches.

DigitalGlobe

DigitalGlobe's WorldView-2 satellite provides high resolution commercial satellite imagery with 0.46 m spatial resolution (panchromatic only). The 0.46 meters resolution of WorldView-2's panchromatic images allows the satellite to distinguish between objects on the ground that are at least 46 cm apart. Similarly DigitalGlobe's QuickBird satellite provides 0.6 meter resolution (at NADIR) panchromatic images.

DigitalGlobe's WorldView-3 satellite provides high resolution commercial satellite imagery with 0.31 m spatial resolution. WVIII also carries a short wave infrared sensor and an atmospheric sensor.

Spot Image

SPOT image of Bratislava.

The 3 SPOT satellites in orbit (Spot 5, 6, 7) provide very high resolution images – 1.5 m for Panchromatic channel, 6m for Multi-spectral (R,G,B,NIR). Spot Image also distributes multiresolution data from other optical satellites, in particular from Formosat-2 (Taiwan) and Kompsat-2 (South Korea) and from radar satellites (TerraSar-X, ERS, Envisat, Radarsat). Spot Image is also the exclusive distributor of data from the high resolution Pleiades satellites with a resolution of 0.50 meter or about 20 inches. The launches occurred in 2011 and 2012, respectively. The company also offers infrastructures for receiving and processing, as well as added value options.

BlackBridge

BlackBridge, previously known as RapidEye, operates a constellation of five satellites, launched in August 2008, the RapidEye constellation contains identical multispectral sensors which are equally calibrated. Therefore, an image from one satellite will be equivalent to an image from any of the other four, allowing for a large amount of imagery to be collected (4 million km² per day), and daily revisit to an area. Each travel on the same orbital plane at 630 km, and deliver images in 5 meter pixel size. RapidEye satellite imagery is especially suited for agricultural, environmental, cartographic and disaster management applications. The company not only offers their imagery, but consults their customers to create services and solutions based on analysis of this imagery.

Image Sat International

Earth Resource Observation Satellites, better known as "EROS" satellites, are lightweight, low earth orbiting, high-resolution satellites designed for fast maneuvering between imaging targets. In the commercial high-resolution satellite market, EROS is the smallest very high resolution satellite; it is very agile and thus enables very high performances. The satellites are deployed in a circular sun-synchronous near polar orbit at an altitude of 510 km (+/- 40 km). EROS satellites imagery applications are primarily for intelligence, homeland security and national development purposes but also employed in a wide range of civilian applications, including: mapping, border control, infrastructure planning, agricultural monitoring, environmental monitoring, disaster response, training and simulations, etc.

EROS A – a high resolution satellite with 1.9–1.2m resolution panchromatic was launched on December 5, 2000.

EROS B – the second generation of Very High Resolution satellites with 70 cm resolution panchromatic, was launched on April 25, 2006.

Imagery Analysis using Artificial Intelligence

Advancements in artificial intelligence have made autonomous, large scale analysis of imagery possible. AI has been taught to process Satellite Imagery with a small degree of error. Studies have used AI to differentiate between different forest types and AI can tell the difference between certain soil and vegetation types. Researchers are using AI to monitor Satellite Imagery for vineyard and grape health as well as having AI estimate wheat harvest size. Projects like SpaceKnow uses AI to conduct case studies in near real-time of deforestation due to wildfires in California and manufacturing activity in China.

As the technology advances, clearer imagery and faster neural networks has allowed for the study of Above Ground Biomass (AGB). This ABG index can describe the size and density of vegetation which scientists use to estimate carbon output and footprints in certain areas. Scientists are eager to apply this data to the study of global warming and climate change. Researchers are developing AI that can monitor refugee movements in war-torn countries, monitor deforestation in the Amazon rain-forest, and show algae blooms in places like the Gulf of Mexico and the Red Sea. Upcoming studies of contaminated surface water and chemical runoff from Fracking are also being planned.

Disadvantages

Because the total area of the land on Earth is so large and because resolution is relatively high, satellite databases are huge and image processing (creating useful images from the raw data) is time-consuming. Preprocessing, such as image destriping is often required. Depending on the sensor used, weather conditions can affect image quality: for example, it is difficult to obtain images for areas of frequent cloud cover such as mountain-tops. For such reasons, publicly available satellite image datasets are typically processed for visual or scientific commercial use by third parties.

Commercial satellite companies do not place their imagery into the public domain and do not sell their imagery; instead, one must be licensed to use their imagery. Thus, the ability to legally make derivative products from commercial satellite imagery is minimized.

Privacy concerns have been brought up by some who wish not to have their property shown from above.

Digital Elevation Model

A digital elevation model (DEM) is a 3D CG representation of a terrain's surface – commonly of a planet (e.g. Earth), moon, or asteroid – created from a terrain's elevation data. A "global DEM" refers to a discrete global grid.

DEMs are used often in geographic information systems, and are the most common basis for digitally produced relief maps. While a digital surface model (DSM) may be useful for landscape modeling, city modeling and visualization applications, a digital terrain model (DTM) is often required for flood or drainage modeling, land-use studies, geological applications, and other applications, and in planetary science.

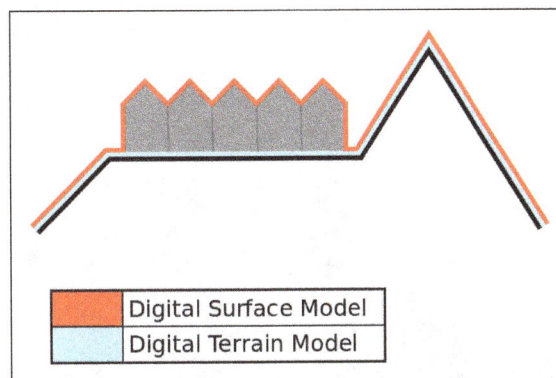

Surfaces represented by a Digital Surface Model include buildings and other objects. Digital Terrain Models represent the bare ground.

There is no universal usage of the terms digital elevation model (DEM), digital terrain model (DTM) and digital surface model (DSM) in scientific literature. In most cases the term digital surface model represents the earth's surface and includes all objects on it. In contrast to a DSM, the digital terrain model (DTM) represents the bare ground surface without any objects like plants and buildings.

DEM is often used as a generic term for DSMs and DTMs, only representing height information without any further definition about the surface. Other definitions equalise the terms DEM and DTM, equalise the terms DEM and DSM, define the DEM as a subset of the DTM, which also represents other morphological elements, or define a DEM as a rectangular grid and a DTM as a three-dimensional model (TIN). Most of the data providers (USGS, ERSDAC, CGIAR, Spot Image) use the term DEM as a generic term for DSMs and DTMs. All datasets which are captured with satellites, airplanes or other flying platforms are originally DSMs (like SRTM or the ASTER GDEM). It is possible to compute a DTM from high resolution DSM datasets with complex algorithms. In the following, the term DEM is used as a generic term for DSMs and DTMs.

Types

Heightmap of Earth's surface (including water and ice), rendered as an equirectangular projection with elevations indicated as normalized 8-bit grayscale, where lighter values indicate higher elevation.

A DEM can be represented as a raster (a grid of squares, also known as a heightmap when representing elevation) or as a vector-based triangular irregular network (TIN). The TIN DEM dataset is also referred to as a primary (measured) DEM, whereas the Raster DEM is referred to as a secondary (computed) DEM. The DEM could be acquired through techniques such as photogrammetry, lidar, IfSAR, land surveying, etc.

DEMs are commonly built using data collected using remote sensing techniques, but they may also be built from land surveying.

Rendering

Relief map of the Sierra Nevada, showing use of both shading and false color as visualization tools to indicate elevation.

The digital elevation model itself consists of a matrix of numbers, but the data from a DEM is often rendered in visual form to make it understandable to humans. This visualization may be in the form of a contoured topographic map, or could use shading and false color assignment (or "pseudo-color") to render elevations as colors (for example, using green for the lowest elevations, shading to red, with white for the highest elevation).

Visualizations are sometime also done as oblique views, reconstructing a synthetic visual image of the terrain as it would appear looking down at an angle. In these oblique visualizations, elevations are sometimes scaled using "vertical exaggeration" in order to make subtle elevation differences more noticeable. Some scientists, however, object to vertical exaggeration as misleading the viewer about the true landscape.

Production

Mappers may prepare digital elevation models in a number of ways, but they frequently use remote sensing rather than direct survey data.

Older methods of generating DEMs often involve interpolating digital contour maps that may have been produced by direct survey of the land surface. This method is still used in mountain areas, where interferometry is not always satisfactory. Note that contour line data or any other sampled elevation datasets (by GPS or ground survey) are not DEMs, but may be considered digital terrain models. A DEM implies that elevation is available continuously at each location in the study area.

Satellite Mapping

One powerful technique for generating digital elevation models is interferometric synthetic aperture radar where two passes of a radar satellite (such as RADARSAT-1 or TerraSAR-X or Cosmo SkyMed), or a single pass if the satellite is equipped with two antennas (like the SRTM instrumentation), collect sufficient data to generate a digital elevation map tens of kilometers on a side with a resolution of around ten meters. Other kinds of stereoscopic pairs can be employed using the digital image correlation method, where two optical images are acquired with different angles taken from the same pass of an airplane or an Earth Observation Satellite (such as the HRS instrument of SPOT5 or the VNIR band of ASTER).

The SPOT 1 satellite provided the first usable elevation data for a sizeable portion of the planet's landmass, using two-pass stereoscopic correlation. Later, further data were provided by the European Remote-Sensing Satellite using the same method, the Shuttle Radar Topography Mission using single-pass SAR and the Advanced Spaceborne Thermal Emission and Reflection Radiometer instrumentation on the Terra satellite using double-pass stereo pairs. The HRS instrument on SPOT 5 has acquired over 100 million square kilometers of stereo pairs.

Planetary Mapping

A tool of increasing value in planetary science has been use of orbital altimetry used to make digital elevation map of planets. A primary tool for this is laser altimetry. Planetary digital elevation maps made using laser altimetry include the Mars Orbiter Laser Altimeter (MOLA) mapping of Mars, the Lunar Orbital Laser Altimeter (LOLA) and Lunar Altimeter (LALT) mapping of the Moon, and the Mercury Laser Altimeter (MLA) mapping of Mercury.

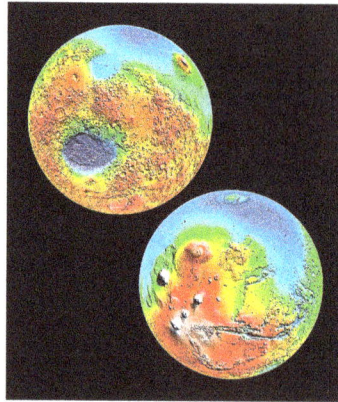

MOLA digital elevation model showing the two hemispheres of Mars.
This image appeared on the cover of *Science* magazine.

Methods for Obtaining Elevation Data used to Create DEMs

Gatewing X100 unmanned aerial vehicle.

- Lidar,

- Radar,

- Stereo photogrammetry from aerial surveys,

- Structure from motion/Multi-view stereo applied to aerial photography,

- Block adjustment from optical satellite imagery,

- Interferometry from radar data,

- Real Time Kinematic GPS,

- Topographic maps,

- Theodolite or total station,

- Doppler radar,

- Focus variation,

- Inertial surveys,

- Surveying and mapping drones,

- Range imaging.

Accuracy

The quality of a DEM is a measure of how accurate elevation is at each pixel (absolute accuracy) and how accurately is the morphology presented (relative accuracy). Several factors play an important role for quality of DEM-derived products:

- Terrain Roughness;

- Sampling Density (Elevation Data Collection Method);

- Grid Resolution Or Pixel Size;

- Interpolation Algorithm;

- Vertical Resolution;

- Terrain Analysis Algorithm;

- Reference 3D products include quality masks that give information on the coastline, lake, snow, clouds, correlation etc.

Uses

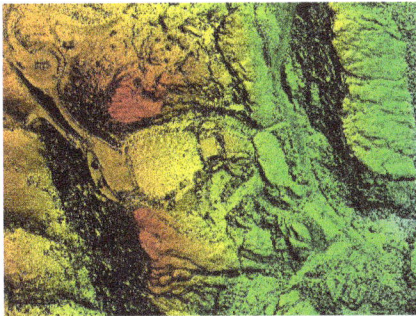

Digital Elevation Model - Red Rocks Amphitheater, Colorado obtained using an UAV (DroneMapper).

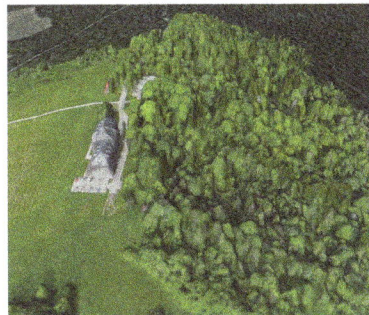

Bezmiechowa airfield 3D Digital Surface Model obtained using Pteryx UAV flying 200 m above hilltop.

Digital Surface Model of motorway interchange construction site. Note that tunnels are closed.

Common uses of DEMs include:

- Extracting terrain parameters for geomorphology,

- Modeling water flow for hydrology or mass movement (for example avalanches and landslides),

- Modeling soils wetness with Cartographic Depth to Water Indexes (DTW-index),

- Creation of relief maps,

- Rendering of 3D visualizations,

- 3D flight planning and TERCOM,

- Creation of physical models (including raised relief maps),

- Rectification of aerial photography or satellite imagery,

- Reduction (terrain correction) of gravity measurements (gravimetry, physical geodesy),

- Terrain analysis in geomorphology and physical geography,

- Geographic information systems (GIS),

- Engineering and infrastructure design,

- Satellite navigation (for example GPS and GLONASS),

- Line-of-sight analysis,

- Base mapping,

- Flight simulation,

- Precision farming and forestry,

- Surface analysis,

- Intelligent transportation systems (ITS),

- Auto safety/advanced driver-assistance systems (ADAS),

- Archaeology.

Example DEM flown with the Gatewing
X100 in Assenede.

Digital Terrain Model Generator + Tex-
tures(Maps) + Vectors.

Global Relief Model

A global relief model, sometimes also denoted as global topography model or composite model, combines digital elevation model (DEM) data over land with digital bathymetry model (DBM) data over water-covered areas (oceans, lakes) to describe Earth's relief. A relief model thus shows how Earth's surface would look like in the absence of water or ice masses.

The relief is represented by a set of heights (elevations or depths) that refer to some height reference surface, often the mean sea level or the geoid. Global relief models are used for a variety of applications including geovisualization, geologic, geomorphologic and geophysical analyses, gravity field modelling as well as geo-statistics.

Measurement

Global relief models are always based on combinations of data sets from different remote sensing techniques. This is because there is no single remote sensing technique that would allow measurement of the relief both over dry and water-covered areas. Elevation data over land is often obtained from LIDAR or inSAR measurements, while bathymetry is acquired based on SONAR and altimetry. Global relief models may also contain elevations of the bedrock (sub-ice-topography) below the ice shields of Greenland and Antarctica. Ice sheet thickness, mostly measured through ice-penetrating RADAR, is subtracted from the ice surface heights to reveal the bedrock.

Spatial Resolution

While digital elevation models describe Earth's land topography often with 1 to 3 arc-second resolution (e.g., from the SRTM or ASTER missions), the global bathymetry (e.g., SRTM30_PLUS) is known to a much lesser spatial resolution in the kilometre-range. The same holds true for models of the bedrock of Antarctica and Greenland. Therefore, global relief models are often constructed at 1 arc-minute resolution (corresponding to about 1.8 km postings). Some products such as the 30 and 15 arc-second resolution SRTM30_PLUS/ SRTM15_PLUS grids offer higher resolution to adequately represent SONAR depth measurements where available. Although grid cells are spaced at 15 or 30 arc-seconds, when SONAR measurements are unavailable, the resolution is much worse (~20-12 km) depending on factors such as water depth.

Public Data Sets

Data sets produced and released to the public include Earth2014, SRTM30_PLUS and ETOPO1.

Earth

The most recent global relief model is Earth2014, developed at Curtin University (Western Australia) and TU Munich (Germany). Earth2014 provides sets of 1 arc-min resolution global grids of Earth's relief in different representations based on the 2013 releases of bedrock and ice-sheet data over Antarctica (Bedmap2) and Greenland (Greenland Bedrock Topography), the 2013 SRTM_30PLUS bathymetry and 2008 SRTM V4.1 SRTM land topography.

Earth2014 provides five different layers of height data, including Earth's surface (lower interface of the atmosphere), topography and bathymetry of the oceans and major lakes, topography, bathymetry and bedrock, ice-sheet thicknesses and rock-equivalent topography. The Earth2014 global grids are provided as heights relative to the mean sea level, and as planetary radii relative to the centre of Earth, which show the shape of the Earth.

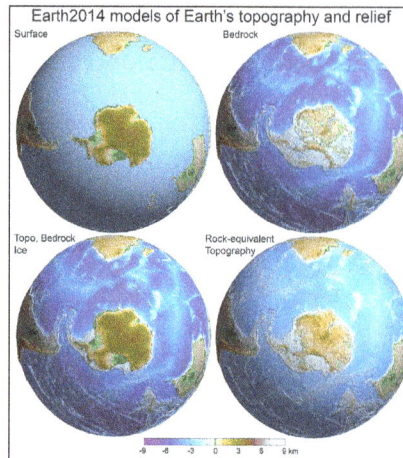

Four different topography layers of the Earth2014 model. Clockwise from
top left: (1) Earth's surface, (2) bedrock, (3) rock-equivalent topography,
(4) bathymetry and ice surface.

SRTM30_PLUS

SRTM30_PLUS is a combined bathymetry and topography model produced by Scripps Institution
of Oceanography (California). The version 15_PLUS comes at 0.25 arc-min resolution (about 450
m postings), while the 30_PLUS version offers 0.5 arc-min (900 m) resolution. The bathymetric
data in SRTM30_PLUS stems from depth soundings (SONAR) and from satellite altimetry. The
bathymetric component of SRTM30_PLUS gets regularly updated with new or improved data sets
in order to continuously improve and refine the description of the sea floor geometry. Over land areas, SRTM30 data from the USGS is included. SRTM30_PLUS provides background information
for Google Earth and Google Maps.

ETOPO1

The ETOPO1 1-arcmin global relief model, produced by the National Geophysical Data Center
(Colorado), provides two layers of relief information. One layer represents the global relief including bedrock over Antarctica and Greenland, and another layer the global relief including ice
surface heights. Both layers include bathymetry over the oceans and some of Earth's major lakes.
ETOPO1 land topography and ocean bathymetry relies on SRTM30 topography and a multitude
of bathymetric surveys that have been merged. Historic versions of ETOPO1 are the ETOPO2 and
ETOPO5 relief models (2 and 5 arc-min resolution).

The ETOPO1 global relief model is based on the 2001 Bedmap1 model of bedrock over Antarctica, which
is now superseded by the significantly improved Bedmap2 bedrock data. The ETOPO1-contained information on ocean depths is superseded through several updates of the SRTM30_PLUS bathymetry.

Geovisualization

Geovisualization or geovisualisation (short for geographic visualization), refers to a set of tools and
techniques supporting the analysis of geospatial data through the use of interactive visualization.

Like the related fields of scientific visualization and information visualization geovisualization emphasizes knowledge construction over knowledge storage or information transmission. To do this, geovisualization communicates geospatial information in ways that, when combined with human understanding, allow for data exploration and decision-making processes.

Traditional, static maps have a limited exploratory capability; the graphical representations are inextricably linked to the geographical information beneath. GIS and geovisualization allow for more interactive maps; including the ability to explore different layers of the map, to zoom in or out, and to change the visual appearance of the map, usually on a computer display. Geovisualization represents a set of cartographic technologies and practices that take advantage of the ability of modern microprocessors to render changes to a map in real time, allowing users to adjust the mapped data on the fly.

Related Fields

Geovisualization is closely related to other visualization fields, such as scientific visualization and information visualization. Owing to its roots in cartography, geovisualization contributes to these other fields by way of the map metaphor, which "has been widely used to visualize non-geographic information in the domains of information visualization and domain knowledge visualization." It is also related to urban simulation.

Applications

Geovisualization has made inroads in a diverse set of real-world situations calling for the decision-making and knowledge creation processes it can provide. The following list provides a summary of some of these applications as they are discussed in the geovisualization literature.

Wildland Fire Fighting

Firefighters have been using sandbox environments to rapidly and physically model topography and fire for wildfire incident command strategic planning. The SimTable is a 3D interactive fire simulator, bringing sandtable exercises to life. The SimTable uses advanced computer simulations to model fires in any area, including local neighborhoods, utilizing actual slope, terrain, wind speed/direction, vegetation, and other factors. SimTable Models were used in Arizona's largest fire on record, the Wallow Fire.

Forestry

Geovisualizers, working with European foresters, used CommonGIS and Visualization Toolkit (VTK) to visualize a large set of spatio-temporal data related to European forests, allowing the data to be explored by non-experts over the Internet. The report summarizing this effort "uncovers a range of fundamental issues relevant to the broad field of geovisualization and information visualization research".

The research team cited the two major problems as the inability of the geovisualizers to convince the foresters of the efficacy of geovisualization in their work and the foresters' misgivings over the dataset's accessibility to non-experts engaging in "uncontrolled exploration". While the geovisualizers focused on the ability of geovisualization to aid in knowledge construction, the foresters preferred the information-communication role of more traditional forms of cartographic representation.

Archaeology

Geovisualization provides archaeologists with a potential technique for mapping unearthed archaeological environments as well as for accessing and exploring archaeological data in three dimensions.

The implications of geovisualization for archaeology are not limited to advances in archaeological theory and exploration but also include the development of new, collaborative relationships between archaeologists and computer scientists.

Environmental Studies

Geovisualization tools provide multiple stakeholders with the ability to make balanced environmental decisions by taking into account "the complex interacting factors that should be taken into account when studying environmental changes". Geovisualization users can use a georeferenced model to explore a complex set of environmental data, interrogating a number of scenarios or policy options to determine a best fit.

Urban Planning

Both planners and the general public can use geovisualization to explore real-world environments and model 'what if' scenarios based on spatio-temporal data. While geovisualization in the preceding fields may be divided into two separate domains—the private domain, in which professionals use geovisualization to explore data and generate hypotheses, and the public domain, in which these professionals present their "visual thinking" to the general public—planning relies more heavily than many other fields on collaboration between the general public and professionals.

Planners use geovisualization as a tool for modeling the environmental interests and policy concerns of the general public. Jiang et al. mention two examples, in which "3D photorealistic representations are used to show urban redevelopment and dynamic computer simulations are used to show possible pollution diffusion over the next few years." The widespread use of the Internet by the general public has implications for these collaborative planning efforts, leading to increased participation by the public while decreasing the amount of time it takes to debate more controversial planning decisions.

Hypsometric Tinting

Hypsometric Tinting (also called layer tinting, elevation tinting, elevation coloring, or hypsometric coloring) is a cartographic technique used to enhance the depiction of relief through placing colors (or tints) between contour lines. Hypsometric tinting can depict ranges of elevation as bands of color in a graduated scheme or as a color ramp applied to the lines themselves. Tints are typically laid semi-transparently over a hillshaded surface.

The first use of color to indicate changes in elevation is attributed to Leonardo da Vinci on a map of Central Italy around 1503. During the late 19th century, Scottish engraver John Bartholomew, Jr. became the first cartographer to color different layers on maps to indicate the height of land or the depth of the sea. He used green to show low ground, brown for higher elevations, and purple fading to white for the highest areas of some maps. The firm where he worked, John Bartholomew and Son, is credited with popularizing this technique, and its color scheme has become conventional in depicting gradual value changes in data. One of the earliest maps to use these ideas was a bathymetric map made by Matthew Fontaine Maury in 1853 in a map to help sailors chart the wind on the sea.

Application

On a map, the different shades represent different elevations. This method of cartographic relief depiction selects colors for the tint which are usually assumed to relate to the ground cover found at the area being mapped.

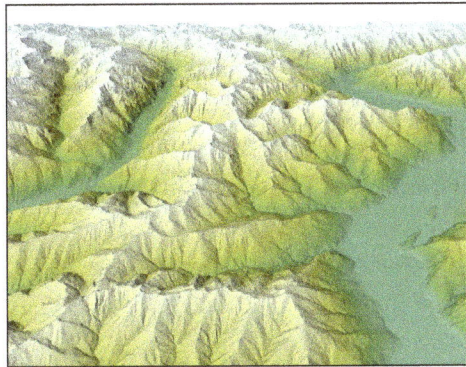

The snow-capped mountains indicate the peaks while the
blue base of the mountain may represent a river valley.

For instance, the highest elevation area on a map might be shaded white to represent snow-capped peaks, while the lowest elevated area may be shaded dark green to represent low-laying valleys or ground level. The advantage of using hypsometric tinting in terrain design is that the map reader can see an accurate depiction of the elevation of any given area on the map.

Another popular use for hypsometric tinting is the display of weather patterns from cold to hot temperatures using warm colors for warmer temperatures and cooler colors for colder temperatures. Often when using hypsometric tinting with elevation, relief shading is added to bring out the elevation. Depending on the audience, maps may contain hypsometric tint simply because the colors are pleasing.

Bathymetric tinting is related to hypsometric tinting. While hypsometric tinting shows land variations, bathymetric tinting portrays the differences in water depth. Showing the depths of a lake is an example of bathymetric tinting.

Application in ArcGIS

There are two ways of depicting hypsometric tinting in ArcGIS. The first way is to use layer tinting as classified values. Classified values allow each elevation to be depicted in areas separated by a neutral gray line. A benefit of this is that colors may be represented exactly how they are entered. However, creating intervals for the elevations along with a color for each can be tedious. The next way is to use a color ramp. This makes transitioning elevations appear smooth and transitional. In ArcMap, the setting for this is called "Stretched". A benefit of using this technique is that the transitional is very natural; however, the user has less control over the colors of the elevations.

Problems

The difficulty in using this method is that the human mind automatically associates certain colors with certain features that may not be completely accurate on the map. For example, many people would automatically associate green on a map with a lush green forest. Dark green is also typically used to show low elevation with hypsometric tinting. This causes problems when a place, such as Death Valley in California, is represented by a dark green because of its low elevation since it does not have very high levels of vegetation growth. This could be resolved by using cross-blended hypsometric tinting. Using this method masks the colors in a way that removes possibly confusing colors but sacrifices some accuracy of the map. It is best used in a small scale map.

References

- Aerial-Survey: gssc.esa.int, Retrieved 15 June, 2019

- Compass-surveying-types-compass-advantages-disadvantages-compass-surveying-example: engineeringcivil.org, Retrieved 16 May, 2019

- Hypsometric-Tinting: wiki-1-1930356585.us-east-1.elb.amazonaws.com, Retrieved 01 February, 2019

- Schowengerdt, Robert A. (2007). Remote sensing: models and methods for image processing (3rd ed.). Academic Press. p. 2. ISBN 978-0-12-369407-2

- Sužiedelytė-Visockienė J, Bagdžiūnaitė R, Malys N, Maliene V (2015). "Close-range photogrammetry enables documentation of environment-induced deformation of architectural heritage". Environmental Engineering and Management Journal. 14 (6): 1371–1381. doi:10.30638/eemj.2015.149

- Gasco, Luis; Asensio, Cesar; de Arcas, Guillermo (2017-05-15). "Communicating airport noise emission data to the general public". Science of the Total Environment. 586: 836–848. doi:10.1016/j.scitotenv.2017.02.063. PMID 28214112

4
Cartography

The science of making maps and charts is referred to as cartography. It helps in communicating spatial, topographic and geographic information of any landform, and terrain. Map projection, cartographic labeling, cartographic generalization, etc. are some of its principles. All these diverse principles of cartography have been carefully analyzed in this chapter.

Map

Map is the graphic representation, drawn to scale and usually on a flat surface, of features—for example, geographical, geological, or geopolitical—of an area of the Earth or of any other celestial body. Globes are maps represented on the surface of a sphere. Cartography is the art and science of making maps and charts.

In order to imply the elements of accurate relationships, and some formal method of projecting the spherical subject to a map plane, further qualifications might be applied to the definition. The tedious and somewhat abstract statements resulting from attempts to formulate precise definitions of maps and charts are more likely to confuse than to clarify. The words map, chart, and plat are used somewhat interchangeably. The connotations of use, however, are distinctive: charts for navigation purposes (nautical and aeronautical), plats (in a property-boundary sense) for land-line references and ownership, and maps for general reference.

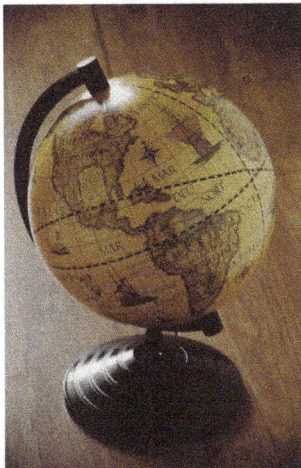
Globe

Cartography is allied with geography in its concern with the broader aspects of the Earth and its life. In early times cartographic efforts were more artistic than scientific and factual. As man explored and recorded his environment, the quality of his maps and charts improved.

Topographic maps are graphic representations of natural and man-made features of parts of the Earth's surface plotted to scale. They show the shape of land and record elevations above sea level, lakes, streams and other hydrographic features, and roads and other works of man. In short, they provide a complete inventory of the terrain and important information for all activities involving the use and development of the land. They provide the bases for specialized maps and data for compilation of generalized maps of smaller scale.

Nautical charts are maps of coastal and marine areas, providing information for navigation. They include depth curves or soundings or both; aids to navigation such as buoys, channel markers, and lights; islands, rocks, wrecks, reefs and other hazards; and significant features of the coastal areas, including promontories, church steeples, water towers, and other features helpful in determining positions from offshore.

The terms hydrography and hydrographer date from the mid-16th century; their focus has become restricted to studies of ocean depths and of the directions and intensities of oceanic currents; though at various times they embraced much of the sciences now called hydrology and oceanography. The British East India Company employed hydrographers in the 18th century, and the first hydrographer of the Royal Navy, Alexander Dalrymple, was appointed in 1795. A naval observatory and hydrographic office was established administratively in the United States Navy in 1854. In 1866 a hydrographic office was established by statute, and in 1962 it was renamed the U.S. Naval Oceanographic Office.

Interest in the charting of oceanic areas away from seacoasts developed in the second half of the 19th century, concurrently with the perfection of submarine cables. As knowledge of the configuration of the ocean basins increased, the attention of scientists was drawn to this field of study. A feature of marine science since the 1950s has been increasingly detailed bathymetric (water-depth measurement) surveys of selected portions of the seafloor. Together with collection of associated geophysical data and sampling of sediments, these studies assist in interpreting the geologic history of the ocean-covered portion of the Earth's crust.

Aeronautical charts provide essential data for the pilot and air navigator. They are, in effect, small-scale topographic maps on which current information on aids to navigation have been superimposed. To facilitate rapid recognition and orientation, principal features of the land that would be visible from an aircraft in flight are shown to the exclusion of less important details.

Mapmaking

Elements

Map design is a twofold process: (1) the determination of user requirements, with attendant decisions as to map content and detail, and (2) the arrangement of content, involving publication scale, standards of treatment, symbolizations, colours, style, and other factors. To some extent user requirements obviously affect standards of treatment, such as publication scale. Otherwise, the latter elements are largely determined on the basis of efficiency, legibility, aesthetic considerations, and traditional practices.

In earlier productions by individual cartographers or small groups, personal judgments determined the nature of the end product, usually with due respect for conventional standards. Map design for large programs, such as the various national map series of today, is quite formal by comparison. In most countries, the requirements of official as well as private users are carefully studied, in conjunction with costs and related factors, when considering possible changes or additions to the current standards.

Requirements of military agencies often have a decisive influence on map design, since it is desirable to avoid the expense of maintaining both civil and military editions of maps. International organizations and committees are additional factors in determining map design. The fact that development of changes in design and content of national map series may become rather involved induces some reluctance to change, as does the fact that map stocks are usually printed in quantities intended to last for 10 or more years. Also, frequent changes in treatments result in extensive overhauls at reprint time, with consequent inconsistencies among the standing editions.

Planning for the production of a national series involves both technical and program considerations. Technical planning involves the choice of a contour interval (the elevation separating adjacent contour lines, or lines of constant elevation), which in turn determines the height of aerial photography and other technical specifications for each project. The sequence of mapping steps, or operational phases, is determined by the overall technical procedures that have been established to achieve the most efficiency.

The program aspects of planning involve fiscal allotments, priorities, schedules, and related matters.

Production controls also play important roles in large programs, where schedules must be balanced with capacities available in the respective phases to avoid backlogs or dormant periods between the mapping steps. Considering that topographic maps may require two years or more to complete, from authorization to final printing, the importance of careful planning is evident. Many factors, including the weather, can converge to cause delays.

Map Scales and Classifications

Map scale refers to the size of the representation on the map as compared to the size of the object on the ground. The scale generally used in architectural drawings, for example, is 1/4 inch to one foot, which means that 1/4 of an inch on the drawing equals one foot on the building being drawn. The scales of models of buildings, railroads, and other objects may be one inch to several feet. Maps cover more extensive areas, and it is usually convenient to express the scale by a representative fraction or proportion, as 1/63,360, 1:63,360, or "one-inch-to-one-mile." The scale of a map is smaller than that of another map when its scale denominator is larger: thus, 1:1,000,000 is a smaller scale than 1:100,000. Most maps carry linear, or bar, scales in one or more margins or in the title blocks.

Nautical charts are constructed on widely different scales and can be generally classified as follows: ocean sailing charts are small-scale charts, 1:5,000,000 or smaller, used for planning long voyages or marking the daily progress of a ship. Sailing charts, used for offshore navigation, show a generalized shoreline, only offshore soundings, and are at a scale between 1:600,000 and 1:5,000,000. As an illustration of chart use, a 10-knot ship covers about 29 inches (74 centimetres) at 1:600,000 scale in a day.

General charts are used for coastwise navigation outside outlying reefs and shoals and are at a scale between 1:100,000 and 1:600,000. Coast charts are intended for use in leaving and entering port or navigating inside outlying reefs or shoals and are at a scale between 1:50,000 and 1:100,000. Harbour charts are for use in harbours and small waterways, with a scale usually larger than 1:50,000.

In rare instances reference may be made to the areal scale of a map, as opposed to the more common linear scale. In such cases the denominator of the fractional reference would be the square of the denominator of the linear scale.

The linear scale may vary within a single map, particularly if the scale is small. Variations in the scale of a map because of the sphericity of the surface it represents may, for practical purposes, be considered as nil. On maps of very large scale, such as 1:24,000, such distortions are negligible (considerably less than variations in the paper from fluctuations of humidity). Precise measurements for engineering purposes are usually restricted to maps of that scale or larger. As maps descend in scale, and distortions inherent to their projection of the spherical surface increase, less accurate measurements of distances may be expected.

Maps may be classified according to scale, content, or derivation. The latter refers to whether a map represents an original survey or has been derived from other maps or source data. Some contain both original and derived elements, usually explained in their footnotes. Producing agencies, technical committees, and international organizations have variously classed maps as large, medium, or small scale. In general, large scale means inch-to-mile and larger, small scale, 1:1,000,000 and smaller, leaving the intermediate field as medium scale. As with most relative terms, these can occasionally lead to confusions but are useful as one practical way to classify maps.

The nature of a map's content, as well as its purpose, provides a primary basis of classification. The terms aeronautical chart, geologic, soil, forest, road, and weather map make obvious their respective contents and purposes. Maps are therefore often classified by the primary purposes they serve. Topographic maps usually form the background for geologic, soil, and similar thematic maps and provide primary elements of the bases upon which many other kinds of maps are compiled.

Geographic and Plane Coordinate Systems

The standard geographic coordinate system of the world involves latitudes north or south of the Equator and longitudes east or west of the Prime Reference Meridian of Greenwich. Map and control point references are stated in degrees, minutes, and seconds carried to the number of decimal places commensurate with the accuracy to which locations have been established.

Geodetic surveys, being of extensive areas, must be adjusted for the Earth's curvature, and reductions must be made to mean sea level for scale. The computations are therefore somewhat involved. As a convenience for engineers and surveyors, many countries have established official plane coordinate systems for each province, state, or sector thereof. By this means, all surveys can be "tied" to control points in the system without transposition to geographic coordinates.

In large countries such as the United States, two basic projections are commonly selected to provide systems with minimum distortions for each state or region. For those long in north–south dimension, the Transverse Mercator is generally used, while for those long in east–west direction, the Lambert conformal (intersecting cone) projection is usually employed. In the case of large regions, two or more zones may be established to limit distortions. Positions of geodetic control points have been computed on the plane coordinate systems and have been made available in published lists.

Basic Data for Compilation

Maps may be compiled from other maps, usually of larger scale, or may be produced from original surveys and photogrammetric compilations. The former are sometimes referred to as derived maps and may include information from various sources, in addition to the maps from which they are principally drawn. Most small-scale series, such as the International Map of the World and World Aeronautical Charts, are compiled from existing information, though new data are occasionally produced to strengthen areas for which little or doubtful information exists. Thus compiled maps may contain fragments of original information while those representing original surveys may include some existing data of higher order, such as details from a city plat.

Road maps, produced by the millions, are compiled from road surveys, topographic maps, and aerial photography. City maps often represent original surveys, made principally to control engineering plans and construction. Some are, however, compiled from enlargements of topographic maps of the area.

Notations regarding the sources from which they were drawn are usually carried on compiled maps. This sometimes includes a reliability diagram showing the areas for which good information was available and those that may be less dependable. Comments regarding certain features or areas, which the editor may deem helpful to the user, may be made in the map itself.

Maps reflecting original surveys, such as a national topographic map series, carry standard marginal information. Date of aerial photography, process and instrumentation employed, notes regarding control and projection, date of field edit, and other information may be included. References to the availability of adjoining maps and those of other scales or series may also be included. Marginal ticks for intervals of plane coordinate systems, military grids, and other reference features are also shown and appropriately labeled.

Symbolization

Symbols are the graphic language of maps and charts that has evolved through generations of cartographers. The symbols doubtless had their origins as simple pictograms that gradually developed into the conventions now generally used.

Early cartographers recognized that common usages and conventions would minimize confusion and to some extent simplify compilation and engraving. Efforts in this direction were made over the years, but cartographers, being artists of a sort, preferred to vary their styles, and effective standardization was not achieved until comparatively recent times. National agencies in most countries established conventions with due regard to practices in other countries. International Map of the World agreements, NATO conventions, and the efforts of the United Nations and of international technical societies aid standardization.

Symbols may be broadly classed as planimetric or hypsographic or may be grouped according to the colours in which they are conventionally printed. Black is used for names and culture, or works of man; blue for water features, or hydrography; brown for relief, or hypsography; green for vegetation classifications; and red for road classes and special information. There are variations, however, particularly in special-purpose series, such as soil and geologic maps. Symbols will also vary, perforce, because of limitations of space in the smaller scales and the feasibility of drawing some features to true scale on large maps. Legends explain the less obvious symbols on many maps, while explanatory sheets or booklets are available for most standard series, providing general data as well as symbol information. When less familiar symbols are used on maps they are often labeled to prevent misunderstanding. The general located-object symbol, with label, is often used in preference to specific symbols for such objects as windmills and lookout towers for similar reasons.

Planimetric features (those shown in "plan," such as streams, shorelines, and roads) are easier to portray than shapes of land and heights above sea level. Mountains were shown on early maps by sketchy lines simulating profile or perspective appearance as envisioned by the cartographer. Little effort was made at true depiction as this was beyond the scope of available information and existing capabilities. Form lines and hachures, among other devices, were also used in attempting to show the land's shape. Hachures are short lines laid down in a pattern to indicate direction of slope. When it became feasible to map rough terrain in more detail, hachuring developed into an artistic speciality. Some hachured maps are remarkable for their detail and fidelity, but much of their quality depends on the skill of draftsman or engraver. They are little used now, except where relief is incidental.

Contours are by far the most common and satisfactory means of showing relief. Contours are lines that connect points of equal elevation. The shorelines of lakes and of the sea are contours. Such lines were little used until the mid-19th century, mainly because surveys had not generally been made in sufficient detail for them to be employed successfully. Mean sea level is the datum to which elevations and contour intervals are generally referred. If mean sea level were to rise 20 feet (six metres) the new shoreline would be where the 20-foot contour line is now shown (assuming that all maps on which it is delineated are reasonably accurate).

The quality of contour maps, until recent times, depended largely on the sketching skill of the topographer. In earlier days funds available for topographic mapping were limited, and not much time could be spent in accurate placement. Later, the accurate location of more control points became feasible. An approximate scale of reliability is therefore indicated by the date of a topographic survey, taking into account the respective situations that existed in various countries. Modern surveys, being based on aerial photos and accurate plotting instruments, are generally better in detail and accuracy than earlier surveys. The personal skill of individual topographers, long a factor in map evaluations, has therefore been substantially eliminated.

Hill shading, or shaded relief, layer or altitude tinting, and special manipulations of contouring are other methods of indicating relief. Hill shading requires considerable artistry, as well as the ability to visualize shapes and interpret contours. For a satisfactory result, background contours are a necessary guide to the artist. Hypsographic tinting is relatively easy, particularly since pho-tomechanical etching and other steps can be used to provide negatives for the respective elevation layers. Difficulty in the reproduction process is sometimes a deterrent to the use of treatments involving the manipulation of contours.

In the past, three-dimensional maps were laboriously constructed for studies in military tactics and for many other purposes. They were costly to produce, as contour layers had to be cut and assembled, filled with plaster and painted, after which streams, roads, etc., had to be drawn on the surface. Lettering then was applied, and models of large structures, such as buildings and bridges, were added. In view of the time and cost involved in such productions, they were sparingly used until recent years when better production methods and materials became available. During and after World War I a process was developed and improved whereby an aluminum sheet was "raised" by tapping along the contours copied on its surface. When the contours selected for tapping were completed, the sheet became, in effect, a mold for shaping plastic sheets to its convolutions. The map was printed on plastic sheets prior to the thermal process of shaping them to the mold. Sets of relief maps were soon produced in this manner for use in schools, military briefings, and many other activities.

During and after World War II the production of plastic relief maps was greatly expanded, while the processes and equipment were further improved and refined. Most significant among these developments was a pantograph-router, which cuts a model from plaster or other suitable material as the selected contours are followed by the operator on a topographic map. This eliminated the distortions inherent in shaping metal sheets by the tapping process. Selected topographic maps are now published in limited relief editions for military instruction, special displays, and general classroom instruction.

Most relief maps are exaggerated severalfold in the vertical scale. The Earth is remarkably smooth, when viewed in actual scale, and many significant features would hardly be distinguishable on a map without some vertical exaggeration. Mt. Everest, for example, is actually only one-seventh of 1 percent of the Earth's radius in height, or only one-third of an inch (about eight millimetres) at a scale of 1:1,000,000. For this reason relief is usually shown at five, or even 10, times actual scale, depending upon the nature of the area represented. This exaggerated relief scale is always explained in the map legend.

Nomenclature

All possible places and features are identified and labeled to maximize the usefulness of the map. Some names must be omitted, particularly from maps of smaller scales, to avoid overcrowding and poor legibility. The editor must decide which names may be eliminated, while arranging place-ments so that a maximum number may be accommodated.

Geographic names are the most important, and sometimes the most troublesome, part of the map nomenclature as a whole. Research on existing maps and related documents for a given area may reveal different names for the same features, variations in spelling, or ambiguous ap-plications of names. The field engineer often finds that local usage is confused and sometimes controversial. Various types of official organizations have been established to study the problems submitted and decide the forms and applications that are to be used in government maps and documents. This function is exercised in the United States by the Board on Geographic Names and in the United Kingdom by the Permanent Committee on Geographical Names; worldwide these activities are coordinated by the United Nations Conference on the Standardization of Geographical Names.

The science of place-names, or toponymics, has become a significant specialty since World War II, and efforts have been made to establish uniform usages and standards of transliteration throughout the world. Renewed interest in completing the remaining sheets of the International Map of the World, collaborations resulting from military alliances, and efforts of committees of international scientific societies and the United Nations have contributed to these efforts.

At the local levels, however, there are different kinds of problems. The larger scales of most basic topographic map series permit the naming of quite minor hilltops, ridges, streams, and branches, for which designations can be obtained locally. In sparsely settled country few names in actual use may be obtained for minor features, while in other areas inquiries may reveal inconsistencies and confusions in both spelling and application of local names. In some areas, for example, local residents may tend to refer to small streams by the name of the present occupant of the headwater area. The occupants of opposite sides of a mountain sometimes refer to it by different names. In coastal areas the waterman and landsman may use different references for the same features.

A prime opportunity for resolving these problems is presented when a topographic map of an area is prepared for publication. By extensive inquiry and documentation and research of local records and deeds, the appropriate form and application of nearly all names can be determined. Publication and distribution of the map as an official document may then tend to solidify local usage and eliminate the confusions that previously existed.

Lettering is selected by the map editor in styles and sizes appropriate to the respective features and the relative importance of each. For topographic maps and most others that follow conventional practice, four basic styles of lettering are used in the Western world. The Roman style is generally used for place-names, political divisions, titles, and related nomenclature. Italic is used for lakes, streams, and other water features. Gothic styles are usually applied to land features such as mountains, ridges, and valleys. Man-made works such as highways, railroads, and canals are usually labeled in slope Gothic capitals, but other distinctive styles are often used for these, together with descriptive notes.

The relative importance of map features is reflected in the different sizes of lettering selected to label them. The most prominent places and features are usually shown in capitals, while lesser ones are labeled with lowercase lettering. In the labeling of cities, however, uppercase lettering is often reserved for state or province capitals. County seats are also labeled in this manner on topographic maps of the United States. For other towns, where lowercase lettering is in order, the sizes selected reflect their relative importance. The use of hand lettering has been abandoned in favour of words and figures printed by type or by a photographic process onto transparent material that is "floated" onto the compilation and anchored by an adhesive wax backing in the proper place. Compass roses and graphic scales are added in the same manner.

Cartography

Cartography is the art and science of graphically representing a geographical area, usually on a flat surface such as a map or chart. It may involve the superimposition of political, cultural, or other nongeographical divisions onto the representation of a geographical area.

Cartography is an ancient discipline that dates from the prehistoric depiction of hunting and fishing territories. The Babylonians mapped the world in a flattened, disk-shaped form, but Claudius Ptolemaeus (Ptolemy) established the basis for subsequent efforts in the 2nd century CE with his eight-volume work (Guide to Geography) that showed a spherical Earth. Maps produced during the Middle Ages followed Ptolemy's guide, but they used Jerusalem as the central feature and placed East at the top. Those representations are often called T-maps because they show only three continents (Europe, Asia, and Africa), separated by the "T" formed by the Mediterranean Sea and the Nile River. More accurate geographical representation began in the 14th century when portolan (seamen's) charts were compiled for navigation.

The discovery of the New World by Europeans led to the need for new techniques in cartography, particularly for the systematic representation on a flat surface of the features of a curved surface—generally referred to as a projection (e.g., Mercator projection, cylindrical projection, and Lambert conformal projection). During the 17th and 18th centuries there was a vast outpouring of printed maps of ever-increasing accuracy and sophistication. Systematic surveys were undertaken involving triangulation that greatly improved map reliability and precision. Noteworthy among the scientific methods introduced later was the use of the telescope for determining the length of a degree of longitude.

Modern cartography largely involves the use of aerial and, increasingly, satellite photographs as a base for any desired map or chart. The procedures for translating photographic data into maps are governed by the principles of photogrammetry and yield a degree of accuracy previously unattainable. The remarkable improvements in satellite photography since the late 20th century and the general availability on the Internet of satellite images have made possible the creation of Google Earth and other databases that are widely available online. Satellite photography has also been used to create highly detailed maps of features of the Moon and of several planets in our solar system and their satellites. In addition, the use of geographic information systems (GIS) has been indispensible in expanding the scope of cartographic subjects.

Map Projection

A map projection is a systematic transformation of the latitudes and longitudes of locations from the surface of a sphere or an ellipsoid into locations on a plane. Maps cannot be created without map projections. All map projections necessarily distort the surface in some fashion. There is no limit to the number of possible map projections.

Projections are a subject of several pure mathematical fields, including differential geometry, projective geometry, and manifolds. However, "map projection" refers specifically to a cartographic projection.

Depending on the purpose of the map, some distortions are acceptable and others are not; therefore, different map projections exist in order to preserve some properties of the sphere-like body at the expense of other properties.

The surfaces of planetary bodies can be mapped even if they are too irregular to be modeled well with a sphere or ellipsoid.

The Earth and other large celestial bodies are generally better modeled as oblate spheroids, whereas small objects such as asteroids often have irregular shapes. It is better modeled by triaxial ellipsoid or prolated spheroid with small eccentricities. Haumea's shape is a Jacobi ellipsoid, with its major axis twice as long as its minor and with its middle axis one and half times as long as its minor. These other surfaces can be mapped as well. Therefore, more generally, a map projection is any method of "flattening" a continuous curved surface onto a plane.

Carl Friedrich Gauss's Theorema Egregium proved that a sphere's surface cannot be represented on a plane without distortion. The same applies to other reference surfaces used as models for the Earth, such as oblate spheroids, ellipsoids and geoids. Since any map projection is a representation of one of those surfaces on a plane, all map projections distort. Every distinct map projection distorts in a distinct way. The study of map projections is the characterization of these distortions.

Maps can be more useful than globes in many situations: they are more compact and easier to store; they readily accommodate an enormous range of scales; they are viewed easily on computer displays; they can facilitate measuring properties of the region being mapped; they can show larger portions of the Earth's surface at once; and they are cheaper to produce and transport. These useful traits of maps motivate the development of map projections.

Projection is not limited to perspective projections, such as those resulting from casting a shadow on a screen, or the rectilinear image produced by a pinhole camera on a flat film plate. Rather, any mathematical function transforming coordinates from the curved surface to the plane is a projection. Few projections in actual use are perspective.

Metric Properties of Maps

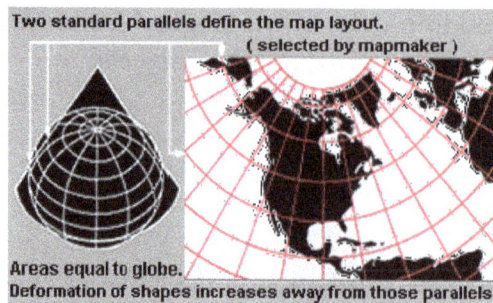

An Albers projection shows areas accurately, but distorts shapes.

Many properties can be measured on the Earth's surface independent of its geography. Some of these properties are:

- Area.
- Shape.

- Direction.

- Bearing.

- Distance.

- Scale.

Map projections can be constructed to preserve at least one of these properties, though only in a limited way for most. Each projection preserves, compromises, or approximates basic metric properties in different ways. The purpose of the map determines which projection should form the base for the map. Because many purposes exist for maps, a diversity of projections have been created to suit those purposes.

Another consideration in the configuration of a projection is its compatibility with data sets to be used on the map. Data sets are geographic information; their collection depends on the chosen datum (model) of the Earth. Different datums assign slightly different coordinates to the same location, so in large scale maps, such as those from national mapping systems, it is important to match the datum to the projection. The slight differences in coordinate assignation between different datums is not a concern for world maps or other vast territories, where such differences get shrunk to imperceptibility.

Distortion

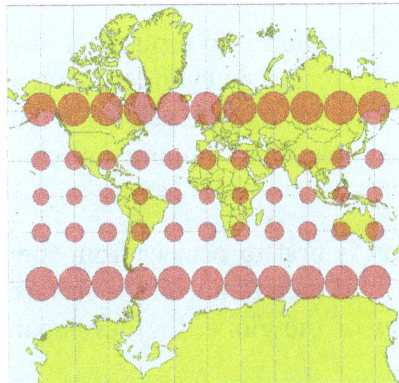

Tissot's Indicatrices on the Mercator projection.

The classical way of showing the distortion inherent in a projection is to use Tissot's indicatrix. For a given point, using the scale factor h along the meridian, the scale factor k along the parallel, and the angle θ' between them, Nicolas Tissot described how to construct an ellipse that characterizes the amount and orientation of the components of distortion. By spacing the ellipses regularly along the meridians and parallels, the network of indicatrices shows how distortion varies across the map.

Design and Construction

The creation of a map projection involves two steps:

- Selection of a model for the shape of the Earth or planetary body (usually choosing between a sphere or ellipsoid). Because the Earth's actual shape is irregular, information is lost in this step.

- Transformation of geographic coordinates (longitude and latitude) to Cartesian (x,y) or polar plane coordinates. In large-scale maps, Cartesian coordinates normally have a simple relation to eastings and northings defined as a grid superimposed on the projection. In small-scale maps, eastings and northings are not meaningful, and grids are not superimposed.

Some of the simplest map projections are literal projections, as obtained by placing a light source at some definite point relative to the globe and projecting its features onto a specified surface. This is not the case for most projections, which are defined only in terms of mathematical formulae that have no direct geometric interpretation. However, picturing the light source-globe model can be helpful in understanding the basic concept of a map projection.

Choosing a Projection Surface

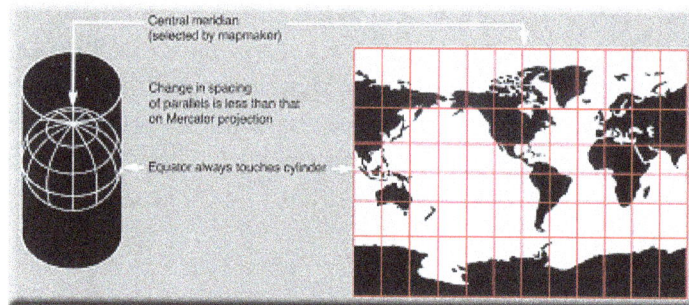

A Miller cylindrical projection maps the globe onto a cylinder.

A surface that can be unfolded or unrolled into a plane or sheet without stretching, tearing or shrinking is called a *developable surface*. The cylinder, cone and the plane are all developable surfaces. The sphere and ellipsoid do not have developable surfaces, so any projection of them onto a plane will have to distort the image. (To compare, one cannot flatten an orange peel without tearing and warping it.)

One way of describing a projection is first to project from the Earth's surface to a developable surface such as a cylinder or cone, and then to unroll the surface into a plane. While the first step inevitably distorts some properties of the globe, the developable surface can then be unfolded without further distortion.

Aspect of the Projection

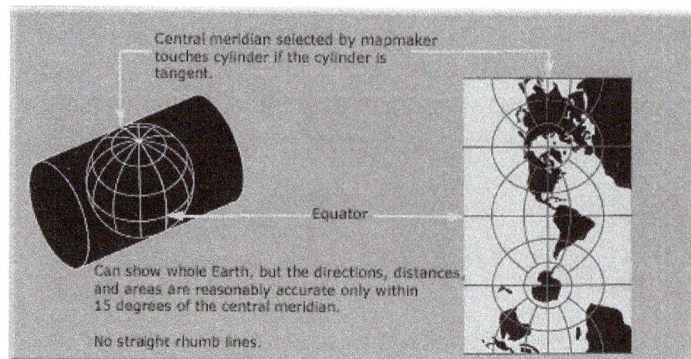

This transverse Mercator projection is mathematically the same as a standard Mercator, but oriented around a different axis.

Once a choice is made between projecting onto a cylinder, cone, or plane, the aspect of the shape must be specified. The aspect describes how the developable surface is placed relative to the globe: it may be normal (such that the surface's axis of symmetry coincides with the Earth's axis), transverse (at right angles to the Earth's axis) or oblique (any angle in between).

Notable Lines

The developable surface may also be either *tangent* or *secant* to the sphere or ellipsoid. Tangent means the surface touches but does not slice through the globe; secant means the surface does slice through the globe. Moving the developable surface away from contact with the globe never preserves or optimizes metric properties, so that possibility is not discussed further here.

Tangent and secant lines (standard lines) are represented undistorted. If these lines are a parallel of latitude, as in conical projections, it is called a standard parallel. The central meridian is the meridian to which the globe is rotated before projecting. The central meridian (usually written λ_o) and a parallel of origin (usually written φ_o) are often used to define the origin of the map projection.

Scale

A globe is the only way to represent the earth with constant scale throughout the entire map in all directions. A map cannot achieve that property for any area, no matter how small. It can, however, achieve constant scale along specific lines.

Some possible properties are:

- The scale depends on location, but not on direction. This is equivalent to preservation of angles, the defining characteristic of a conformal map.

- Scale is constant along any parallel in the direction of the parallel. This applies for any cylindrical or pseudocylindrical projection in normal aspect.

- Combination of the above: the scale depends on latitude only, not on longitude or direction. This applies for the Mercator projection in normal aspect.

- Scale is constant along all straight lines radiating from a particular geographic location. This is the defining characteristic of an equidistant projection such as the Azimuthal equidistant projection. There are also projections (Maurer's Two-point equidistant projection, Close) where true distances from *two* points are preserved.

Choosing a Model for the Shape of the Body

Projection construction is also affected by how the shape of the Earth or planetary body is approximated.

Selecting a model for a shape of the Earth involves choosing between the advantages and disadvantages of a sphere versus an ellipsoid. Spherical models are useful for small-scale maps such as world atlases and globes, since the error at that scale is not usually noticeable or important enough to justify using the more complicated ellipsoid. The ellipsoidal model is commonly used to construct topographic maps and for other large- and medium-scale maps that need to accurately depict the land surface. Auxiliary latitudes are often employed in projecting the ellipsoid.

A third model is the geoid, a more complex and accurate representation of Earth's shape co-incident with what mean sea level would be if there were no winds, tides, or land. Compared to the best fitting ellipsoid, a geoidal model would change the characterization of important properties such as distance, conformality and equivalence. Therefore, in geoidal projections that preserve such properties, the mapped graticule would deviate from a mapped ellipsoid's graticule. Normally the geoid is not used as an Earth model for projections, however, because Earth's shape is very regular, with the undulation of the geoid amounting to less than 100 m from the ellipsoidal model out of the 6.3 million m Earth radius. For irregular planetary bodies such as asteroids, however, sometimes models analogous to the geoid are used to project maps from.

Classification

A fundamental projection classification is based on the type of projection surface onto which the globe is conceptually projected. The projections are described in terms of placing a gigantic surface in contact with the earth, followed by an implied scaling operation. These surfaces are cylindrical (e.g. Mercator), conic (e.g. Albers), and plane (e.g. stereographic). Many mathematical projections, however, do not neatly fit into any of these three conceptual projection methods. Hence other peer categories have been described in the literature, such as pseudoconic, pseudocylindrical, pseudoazimuthal, retroazimuthal, and polyconic.

Another way to classify projections is according to properties of the model they preserve. Some of the more common categories are:

- Preserving direction (azimuthal or zenithal), a trait possible only from one or two points to every other point.

- Preserving shape locally (conformal or orthomorphic).

- Preserving area (equal-area or equiareal or equivalent or authalic).

- Preserving distance (equidistant), a trait possible only between one or two points and every other point.

- Preserving shortest route, a trait preserved only by the gnomonic projection.

Because the sphere is not a developable surface, it is impossible to construct a map projection that is both equal-area and conformal.

Projections by Surface

The three developable surfaces (plane, cylinder, cone) provide useful models for understanding, describing, and developing map projections. However, these models are limited in two fundamental ways. For one thing, most world projections in use do not fall into any of those categories. For another thing, even most projections that do fall into those categories are not naturally attainable through physical projection.

No reference has been made in the above definitions to cylinders, cones or planes. The projections are termed cylindric or conic because they can be regarded as developed on a cylinder or a cone, as the case may be, but it is as well to dispense with picturing cylinders and cones, since they have given rise to much misunderstanding. Particularly is this so with regard to the conic projections with two standard parallels: they may be regarded as developed on cones, but they are cones which bear no simple relationship to the sphere. In reality, cylinders and cones provide us with convenient descriptive terms, but little else.

Lee's objection refers to the way the terms cylindrical, conic, and planar (azimuthal) have been abstracted in the field of map projections. If maps were projected as in light shining through a globe onto a developable surface, then the spacing of parallels would follow a very limited set of possibilities. Such a cylindrical projection (for example) is one which:

1. Is rectangular;

2. Has straight vertical meridians, spaced evenly;

3. Has straight parallels symmetrically placed about the equator;

4. Has parallels constrained to where they fall when light shines through the globe onto the cylinder, with the light source someplace along the line formed by the intersection of the prime meridian with the equator, and the center of the sphere.

(If you rotate the globe before projecting then the parallels and meridians will not necessarily still be straight lines. Rotations are normally ignored for the purpose of classification.)

Where the light source emanates along the line described in this last constraint is what yields the differences between the various "natural" cylindrical projections. But the term *cylindrical* as used in the field of map projections relaxes the last constraint entirely. Instead the parallels can be placed according to any algorithm the designer has decided suits the needs of the map. The famous Mercator projection is one in which the placement of parallels does not arise by "projection"; instead parallels are placed how they need to be in order to satisfy the property that a course of constant bearing is always plotted as a straight line.

Cylindrical

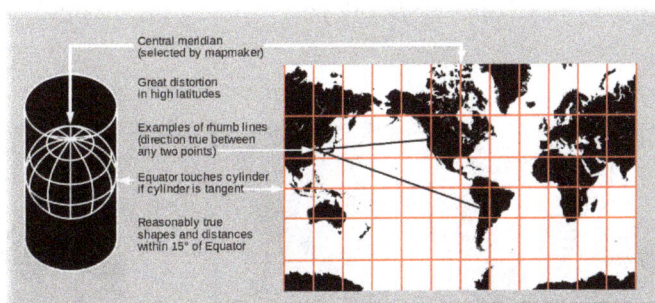

The Mercator projection shows rhumbs as straight lines. A rhumb is a
course of constant bearing. Bearing is the compass direction of movement.

A "normal cylindrical projection" is any projection in which meridians are mapped to equally spaced vertical lines and circles of latitude (parallels) are mapped to horizontal lines.

The mapping of meridians to vertical lines can be visualized by imagining a cylinder whose axis coincides with the Earth's axis of rotation. This cylinder is wrapped around the Earth, projected onto, and then unrolled.

By the geometry of their construction, cylindrical projections stretch distances east-west. The amount of stretch is the same at any chosen latitude on all cylindrical projections, and is given by the secant of the latitude as a multiple of the equator's scale. The various cylindrical projections are distinguished from each other solely by their north-south stretching (where latitude is given by φ):

- North-south stretching equals east-west stretching (sec φ): The east-west scale matches the north-south scale: conformal cylindrical or Mercator; this distorts areas excessively in high latitudes.

- North-south stretching grows with latitude faster than east-west stretching (sec^2 φ): The cylindric perspective (or central cylindrical) projection; unsuitable because distortion is even worse than in the Mercator projection.

- North-south stretching grows with latitude, but less quickly than the east-west stretching: such as the Miller cylindrical projection (sec 4/5φ).

- North-south distances neither stretched nor compressed (1): equirectangular projection or "plate carrée".

- North-south compression equals the cosine of the latitude (the reciprocal of east-west stretching): equal-area cylindrical. This projection has many named specializations differing only in the scaling constant, such as the Gall–Peters or Gall orthographic (undistorted at the 45° parallels), Behrmann (undistorted at the 30° parallels), and Lambert cylindrical equal-area (undistorted at the equator). Since this projection scales north-south distances by the reciprocal of east-west stretching, it preserves area at the expense of shapes.

In the first case (Mercator), the east-west scale always equals the north-south scale. In the second case (central cylindrical), the north-south scale exceeds the east-west scale everywhere away from the equator. Each remaining case has a pair of secant lines—a pair of identical latitudes of opposite sign (or else the equator) at which the east-west scale matches the north-south-scale.

Normal cylindrical projections map the whole Earth as a finite rectangle, except in the first two cases, where the rectangle stretches infinitely tall while retaining constant width.

Pseudocylindrical

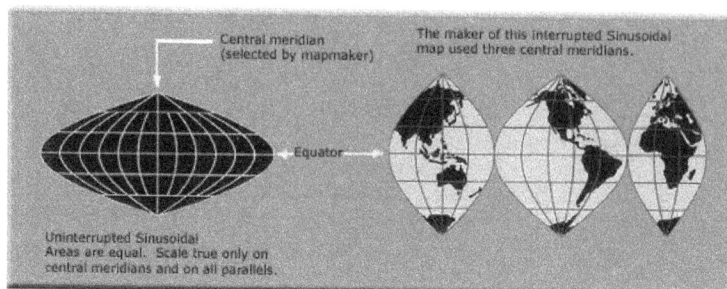

A sinusoidal projection shows relative sizes accurately, but grossly distorts shapes. Distortion can be reduced by "interrupting" the map.

Pseudocylindrical projections represent the *central* meridian as a straight line segment. Other meridians are longer than the central meridian and bow outward, away from the central meridian. Pseudocylindrical projections map parallels as straight lines. Along parallels, each point from the surface is mapped at a distance from the central meridian that is proportional to its difference in longitude from the central meridian. Therefore, meridians are equally spaced along a given parallel. On a pseudocylindrical map, any point further from the equator than some other point has a higher latitude than the other point, preserving north-south relationships. This trait is useful when illustrating phenomena that depend on latitude, such as climate. Examples of pseudocylindrical projections include:

- Sinusoidal, which was the first pseudocylindrical projection developed. On the map, as in reality, the length of each parallel is proportional to the cosine of the latitude. The area of any region is true.

- Collignon projection, which in its most common forms represents each meridian as two straight line segments, one from each pole to the equator.

- Tobler hyperelliptical

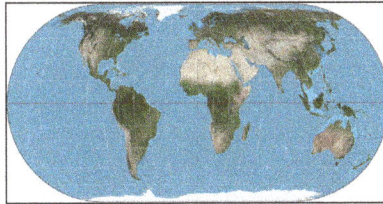

- Mollweide

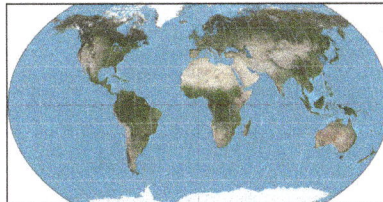

- Goode homolosine

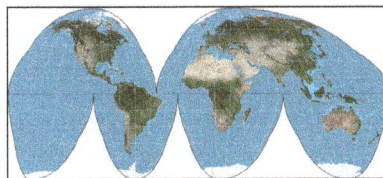

- Eckert IV

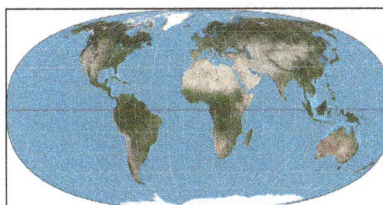

- Eckert VI

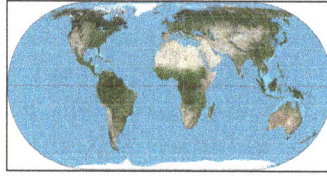

- Kavrayskiy VII

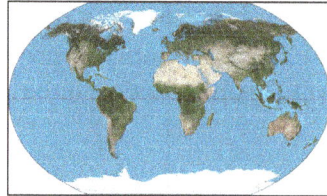

Hybrid

The HEALPix projection combines an equal-area cylindrical projection in equatorial regions with the Collignon projection in polar areas.

Conic

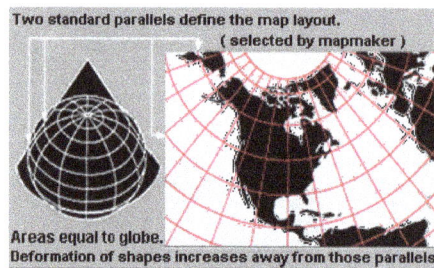
Albers conic.

The term "conic projection" is used to refer to any projection in which meridians are mapped to equally spaced lines radiating out from the apex and circles of latitude (parallels) are mapped to circular arcs centered on the apex.

When making a conic map, the map maker arbitrarily picks two standard parallels. Those standard parallels may be visualized as secant lines where the cone intersects the globe—or, if the map maker chooses the same parallel twice, as the tangent line where the cone is tangent to the globe. The resulting conic map has low distortion in scale, shape, and area near those standard parallels. Distances along the parallels to the north of both standard parallels or to the south of both standard parallels are stretched; distances along parallels between the standard parallels are compressed. When a single standard parallel is used, distances along all other parallels are stretched.

Conic projections that are commonly used are:

- Equidistant conic, which keeps parallels evenly spaced along the meridians to preserve a constant distance scale along each meridian, typically the same or similar scale as along the standard parallels.

- Albers conic, which adjusts the north-south distance between non-standard parallels to compensate for the east-west stretching or compression, giving an equal-area map.

- Lambert conformal conic, which adjusts the north-south distance between non-standard parallels to equal the east-west stretching, giving a conformal map.

Pseudoconic

- Bonne, an equal-area projection on which most meridians and parallels appear as curved lines. It has a configurable standard parallel along which there is no distortion.

- Werner cordiform, upon which distances are correct from one pole, as well as along all parallels.

- American polyconic.

Azimuthal (Projections onto a Plane)

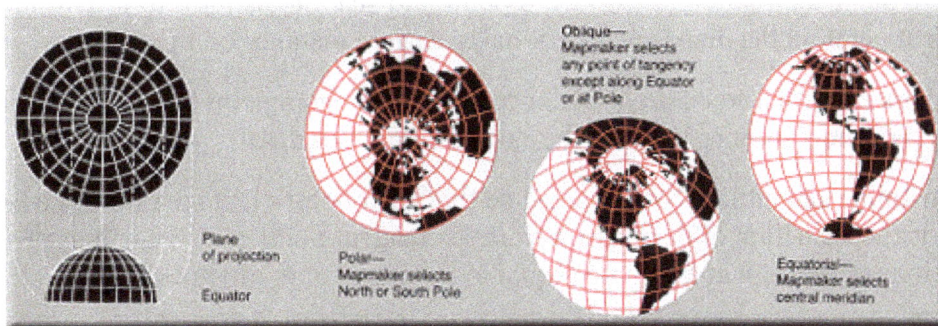

An azimuthal equidistant projection shows distances and directions accurately from the center point, but distorts shapes and sizes elsewhere.

Azimuthal projections have the property that directions from a central point are preserved and therefore great circles through the central point are represented by straight lines on the map. These projections also have radial symmetry in the scales and hence in the distortions: map distances from the central point are computed by a function $r(d)$ of the true distance d, independent of the angle; correspondingly, circles with the central point as center are mapped into circles which have as center the central point on the map.

The mapping of radial lines can be visualized by imagining a plane tangent to the Earth, with the central point as tangent point.

The radial scale is $r'(d)$ and the transverse scale $r(d)/(R \sin d/R)$ where R is the radius of the Earth.

Some azimuthal projections are true perspective projections; that is, they can be constructed mechanically, projecting the surface of the Earth by extending lines from a point of perspective (along an infinite line through the tangent point and the tangent point's antipode) onto the plane:

- The gnomonic projection displays great circles as straight lines. Can be constructed by using a point of perspective at the center of the Earth. $r(d) = c \tan d/R$; so that even just a hemisphere is already infinite in extent.

- The General Perspective projection can be constructed by using a point of perspective outside the earth. Photographs of Earth (such as those from the International Space Station) give this perspective.

- The orthographic projection maps each point on the earth to the closest point on the plane. Can be constructed from a point of perspective an infinite distance from the tangent point; $r(d) = c \sin d/R$. Can display up to a hemisphere on a finite circle. Photographs of Earth from far enough away, such as the Moon, approximate this perspective.

- The stereographic projection, which is conformal, can be constructed by using the tangent point's antipode as the point of perspective $r(d) = c \tan d/2R$; the scale is $c/(2R \cos^2 d/2R)$ can display nearly the entire sphere's surface on a finite circle. The sphere's full surface requires an infinite map.

Other azimuthal projections are not true perspective projections:

- Azimuthal equidistant: $r(d) = cd$; it is used by amateur radio operators to know the direction to point their antennas toward a point and see the distance to it. Distance from the tangent point on the map is proportional to surface distance on the earth.

- Lambert azimuthal equal-area. Distance from the tangent point on the map is proportional to straight-line distance through the earth: $r(d) = c \sin d/2R$.

- Logarithmic azimuthal is constructed so that each point's distance from the center of the map is the logarithm of its distance from the tangent point on the Earth. $r(d) = c \ln d/d_0$); locations closer than at a distance equal to the constant d_0 are not shown.

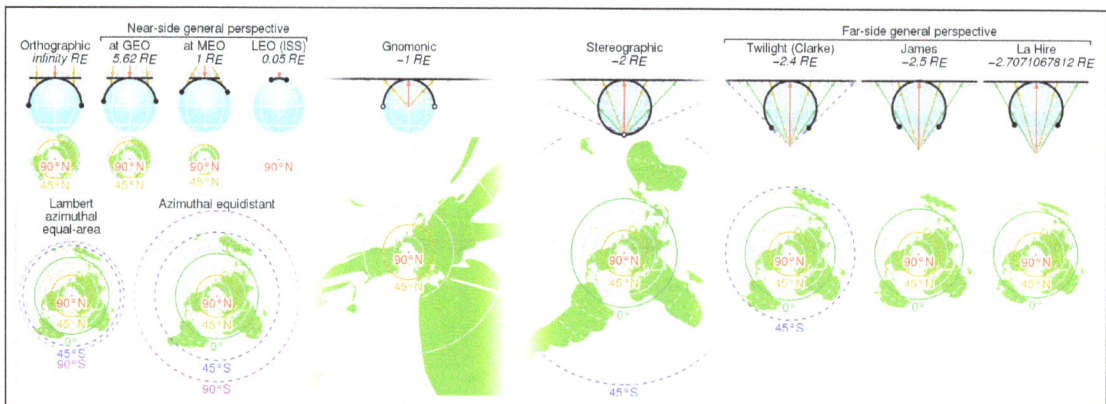

Comparison of some azimuthal projections centred on 90° N at the same scale, ordered by projection altitude in Earth radii. (click for detail).

Projections by Preservation of a Metric Property

Conformal

Conformal, or orthomorphic, map projections preserve angles locally, implying that they map infinitesimal circles of constant size anywhere on the Earth to infinitesimal circles of varying sizes on the map. In contrast, mappings that are not conformal distort most such small circles into ellipses of distortion. An important consequence of conformality is that relative angles at each point of the map are correct, and the local scale (although varying throughout the map) in every direction around any one point is constant. These are some conformal projections:

- Mercator: Rhumb lines are represented by straight segments.

- Transverse Mercator.

- Stereographic: Any circle of a sphere, great and small, maps to a circle or straight line.

- Roussilhe.

- Lambert conformal conic.

- Peirce quincuncial projection.

- Adams hemisphere-in-a-square projection.

- Guyou hemisphere-in-a-square projection.

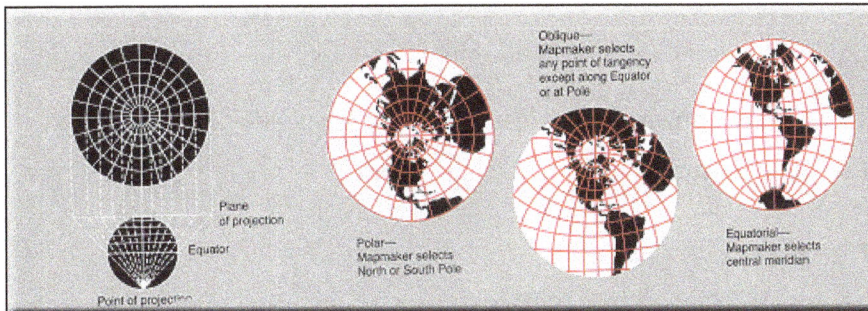

A stereographic projection is conformal and perspective but not equal area or equidistant.

Equal-area

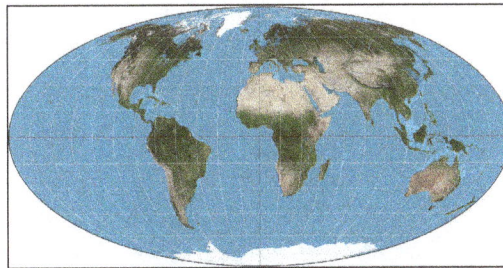

The equal-area Mollweide projection.

Equal-area maps preserve area measure, generally distorting shapes in order to do that. Equal-area maps are also called equivalent or authalic. These are some projections that preserve area:

- Albers conic,

- Bonne,

- Bottomley,

- Collignon,

- Cylindrical equal-area,

- Eckert II, IV and VI,

- Equal Earth,

- Gall orthographic (also known as Gall–Peters, or Peters, projection),

- Goode's homolosine,

- Hammer,

- Hobo–Dyer,

- Lambert azimuthal equal-area,

- Lambert cylindrical equal-area,

- Mollweide,

- Sinusoidal,

- Strebe 1995,

- Snyder's equal-area polyhedral projection, used for geodesic grids,

- Tobler hyperelliptical,

- Werner.

Equidistant

A two-point equidistant projection of Eurasia.

These are some projections that preserve distance from some standard point or line:

- Equirectangular—distances along meridians are conserved.

- Plate carrée—an Equirectangular projection centered at the equator.

- Azimuthal equidistant—distances along great circles radiating from centre are conserved.

- Equidistant conic.

- Sinusoidal—distances along parallels are conserved.

- Werner cordiform distances from the North Pole are correct as are the curved distance on parallels.

- Soldner.

- Two-point equidistant: two "control points" are arbitrarily chosen by the map maker. Distance from any point on the map to each control point is proportional to surface distance on the earth.

Gnomonic

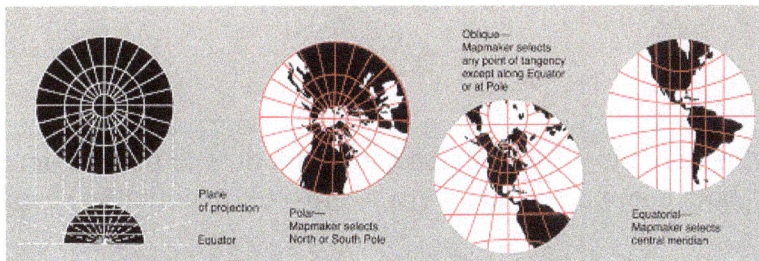

The Gnomonic projection is thought to be the oldest map projection, developed by Thales in the 6th century BC.

Great circles are displayed as straight lines:

- Gnomonic projection.

Retroazimuthal

Direction to a fixed location B (the bearing at the starting location A of the shortest route) corresponds to the direction on the map from A to B:

- Littrow—the only conformal retroazimuthal projection.

- Hammer retroazimuthal—also preserves distance from the central point.

- Craig retroazimuthal *aka* Mecca or Qibla—also has vertical meridians.

Compromise Projections

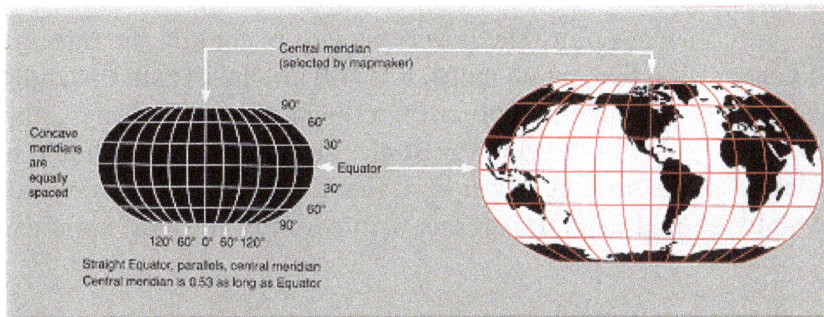

The Robinson projection was adopted by *National Geographic* but abandoned by them for the Winkel tripel.

Compromise projections give up the idea of perfectly preserving metric properties, seeking instead to strike a balance between distortions, or to simply make things "look right". Most of these types of projections distort shape in the polar regions more than at the equator.

Cartographic Labeling

Cartographic labeling is a form of typography and strongly deals with form, style, weight and size of type on a map. Essentially, labeling denotes the correct way to label features (points, arcs, or polygons).

Form

In type, form describes anything from lengths between letters to the case and color of the font. Form works well for both nominal (qualitative) and ordered (quantitative) data.

Italics

Italics describe the sloping of letters setting it apart from non-italicized words (or vice versa). Using italics on a map also slightly decreases the size of the font as it shapely squeezes it around features. When introduced, the idea was to condense the text by italicizing it, thus creating more text on the pages. The slope in the font was created to mimic the flow of cursive handwriting and thus, the angles of italic letters range anywhere from 11 to 30 degree and consequently, serifs are absent.

As a general rule on maps, the smaller the point size of a font, the more condensed and difficult it becomes to read. In an example of labeling a globe, ocean features are generally italicized to give an obvious discernment. In cartographic conventions, natural features are adequate in italics such as the aforementioned hydrographic features.

Case

Case is another way of emphasizing—whether it be uppercase, lowercase or a combination of the two (or even different size points within the same case). In general, uppercase fonts denote a higher emphasis, but according to Bringhurst, an uppercase initial of a word has the seniority; but the lowercase letters have the control. In other words, the strong boldness of a larger letter draws the audience into its viewpoint. The lowercase letters contain the information needed to convey further. When viewing the text on maps, it is still crucial to gain the audience's attention as a way of informing them of something other than the maps. As for design, uppercase is much harder to read than mixed-use. In the globe example, mountain ranges should be in uppercase. When showing a larger scale, such as a region of the United States, it is useful to classify different case sizes. States should be in uppercase, with counties in small uppercase, and cities in lowercase.

Color

Color (value and hue) alterations also allow for further emphasis on certain features. By changing the color of the font to correspond to the feature it is representing, the two become joined. If the cartographer were to label a river, the extra emphasis would be inherent if the font chosen was blue, to correspond with the blue feature (arc). On the contrary, though, this is not always necessarily the case. If the cartographer chose a color of the font for an ocean feature (polygon), blue would not be the obvious choice because it would appear to be washed out and thus, no emphasis. In this case, it is useful to label the feature with a more rich, bolder color (such as the black font on blue polygon).

Spacing

The spacing of the letters on features also gives a more appealing map visually speaking. By enlarging the increments between each letter of a word, the word in turn, becomes more pronounced. In the case for a long arc feature (river), to add more emphasis on the label, the letters would need to be extended or stretched. On the other hand, in some cases, the letters would have to be condensed (shortened increment gaps) to give a more proportional label for a feature.

Style

Serifs

The type style affects to the overall look of the map and is adequately used to symbolize nominal (qualitative) data within the map. In general, style amounts to the use of serifs versus sans serifs. A serif is, by definition, a cross-line at the end of a stroke along with a letter. On a map, the text that is chosen should be consistent. Generally, serif fonts are utilized to give a more regimented block body of text—similar to those used in traditional printing. Serifs are more widely used for historical information or a historical map.

Sans Serifs

The serif counterpart is sans serifs (meaning without serifs). Sans serif fonts are the more modern of the two fonts. But choosing one over the other requires that the audience will be able to read the text without strain. Generally, sans serifs are not for large bodies of text in print but instead, are ideal for the internet. On the same facet, sans serifs are optimal for a more-clean appearance in such places like a header, title, or legend. In map design, it's useful to also use sans serifs for natural features.

Weight

The type weight provides a substantial amount of emphasis of the cartographer's choosing. Weight is important because it involves the difference between bold and regular contrast. The degree of power that is increased with weight, must be proportional to the size of the letter. If not, a letter can be too intense and thus more difficult to read. Similarly, the spacing between the letters must be extended to provide adequate to read smoothly. Bold text creates direct attention to the eyes of the audience to pronounce certain information from cartographer.

Size

The type size of fonts stresses the importance and emphasis of the intended map. Size is expressed in points through the American point system with 1 point equaling 1/72" of vertical height. Furthermore, points also show the spacing between letters, words and lines. A larger size implies more importance or a greater relative quantity; smaller denotes less importance or less quantity. For design purposes, text using a size of less than 6 point is difficult to read. On the contraire, text that is larger than 26 point is too cumbersome for a standard-size paper format. For titles, a font larger than 10 point generally allows for a good working title. Also, it is important to use at least a 2-point difference between type sizes to allow the audience to see subtle changes.

Placement

With all of the type in order and adequately designed, the final step is the correct placement of labels. Placement describes each feature and its subsequent label(s). For area features, it is important to curve and extend the spaces to properly fill in the areas enough that the audience can discern different areas. As a cartographic convention, labels are usually as horizontal as possible with no upside-down labels. For line features, it is useful to allow the label to conform to the line pattern. Similar to a river (e.g. geographic features), the label should flow around the edges along the line being careful not to have the letters too extended. For point patterns, the minor patterns to follow include keeping labels on/in their respective features (e.g. coastal cities with labels on the land and not ocean). The major pattern for points is the placement along the point itself. The most widely accepted pattern is to start at the center and work outward towards the northeast quadrant from the point. Many studies have been researched to address the correct strategy for the placements. The point feature cartographic label placement (PFCLP) problem offers the solutions when point boxes overlap. Many software features automatically choose label placements for the cartographer, but these are not always a fail-safe option. The use of good judgment and cartographic conventions are important to gain the best placement.

Cartographic Generalization

Cartographic Gneralization is the processes of selection and summarizing of the contents in drafting geographical maps. The purpose of such generalization is to preserve and distinguish on a map the main and typical outlines and the characteristic peculiarities of the features shown in accordance with the function, subject, and possible scales of the map. Map scale has the most obvious effect on cartographic generalization. For example, the representation of an area of 1 sq km on a map with a scale of 1:1000 will occupy an area of 1 sq m; with a scale of 1:10,000, 1 sq decimeter; with a scale of 1:100,000, 1 sq cm; and with a scale of 1:1,000,000, 1 sq mm. The depiction of a locality in all of these scales with identical detail and saturation is impossible. The exclusion of details and less important elements is inevitable as the scale becomes smaller. However, the effect of the scale is not only to limit the amount of space available on the map: on a small-scale map, which covers considerable area, details lose their significance and, if retained, would make it more difficult to perceive the main objects on the map. For example, an overall picture of the mountain systems of the Caucasus can be conveyed only on a small-scale and very generalized map, not on detailed topographical maps. Cartographic generalization is influenced by geographical conditions: the same features (or their peculiarities) are evaluated differently for different landscapes or according to the special nature of their relationship to other features—for example, wells are an important element in all topographical maps of desert and semidesert regions, but they are not indicated on similar maps showing areas with good water supply. Cartographic generalization is particularly affected by the function of the map. For example, in the case of a reference map, as much information as possible is provided, whereas on an educational map of the same scale the number of features shown will be reduced and limited to the requirements of the school program.

Cartographic generalization may appear in the following ways: (1) the selection of objects (the restriction of the contents of the map to objects that are essential), (2) the carefully considered simplification of contours (the planned outlining of objects, both linear and those that occupy an area, in which the peculiarities of the outlines typical of such objects are maintained and sometimes even emphasized—the sickle shape of oxbow lakes or the circular shape of lakes on out-wash plains), (3) the generalization of quantitative characteristics by reducing the number of divisions within which quantitative differences for specific features shown on the map are indicated (for example, in the case of a population scale for built-up areas, combining two divisions on the scale, such as "less than 500 inhabitants" and "from 500 to 2,000 inhabitants" into one division, "less than 2,000 inhabitants"), (4) the generalization of qualitative characteristics by simplifying the classifications for the features being shown (not subdividing forests according to type when showing vegetation on topographical maps), and (5) the replacement of individual features by general designations (indicating a population center by blocks and a geometrical sign instead of marking individual buildings).

The establishment of the principles governing cartographic generalization is an important scientific problem of cartography. An example is the establishment of rules of selection in a mathematical form, particularly in the form of quantitative indexes, the criteria that determine the conditions for indicating on the map objects of various categories (for example, the obligation to indicate all cities having 10,000 or more inhabitants). The selection indexes vary according to the map and the geographical region. The development of the mathematical bases of cartographic generalization has acquired considerable importance as a result of the introduction of automation into the processes of drafting and use of maps.

References

- Map, science: britannica.com, Retrieved 16 July, 2019

- Cartography, science: britannica.com, Retrieved 18 June, 2019

- Cartographic-Generalization: encyclopedia2.thefreedictionary.com, Retrieved 05 May, 2019

- Snyder, John P. (1993). Flattening the earth: two thousand years of map projections. University of Chicago Press. ISBN 0-226-76746-9

5

Geographic Information System

Geographic information refers to the data associated with any specific location. A system which collects, stores, analysis and manages such data is termed as geographic information system. Some of its processes are spatial analysis, participatory geographic information system, CyberGIS, etc. This chapter discusses in detail these processes related to geographic information system.

A geographic information system (GIS) is a computer system for capturing, storing, checking, and displaying data related to positions on Earth's surface. By relating seemingly unrelated data, GIS can help individuals and organizations better understand spatial patterns and relationships.

GIS technology is a crucial part of spatial data infrastructure, which the White House defines as "the technology, policies, standards, human resources, and related activities necessary to acquire, process, distribute, use, maintain, and preserve spatial data."

GIS can use any information that includes location. The location can be expressed in many different ways, such as latitude and longitude, address, or ZIP code.

Many different types of information can be compared and contr asted using GIS. The system can include data about people, such as population, income, or education level. It can include information about the landscape, such as the location of streams, different kinds of vegetation, and different kinds of soil. It can include information about the sites of factories, farms, and schools; or storm drains, roads, and electric power lines.

With GIS technology, people can compare the locations of different things in order to discover how they relate to each other. For example, using GIS, a single map could include sites that produce pollution, such as factories, and sites that are sensitive to pollution, such as wetlands and rivers. Such a map would help people determine where water supplies are most at risk.

Data Capture

Data Formats

GIS applications include both hardware and software systems. These applications may include cartographic data, photographic data, digital data, or data in spreadsheets.

Cartographic data are already in map form, and may include such information as the location of rivers, roads, hills, and valleys. Cartographic data may also include survey data, mapping information which can be directly entered into a GIS.

Photographic interpretation is a major part of GIS. Photo interpretation involves analyzing aerial photographs and assessing the features that appear.

Digital data can also be entered into GIS. An example of this kind of information is computer data collected by satellites that show land use—the location of farms, towns, and forests.

Remote sensing provides another tool that can be integrated into a GIS. Remote sensing includes imagery and other data collected from satellites, balloons, and drones.

Finally, GIS can also include data in table or spreadsheet form, such as population demographics. Demographics can range from age, income, and ethnicity to recent purchases and Internet browsing preferences.

GIS technology allows all these different types of information, no matter their source or original format, to be overlaid on top of one another on a single map. GIS uses location as the key index variable to relate these seemingly unrelated data.

Putting information into GIS is called data capture. Data that are already in digital form, such as most tables and images taken by satellites, can simply be uploaded into GIS. Maps, however, must first be scanned, or converted to digital format.

The two major types of GIS file formats are raster and vector. Raster formats are grids of cells or pixels. Raster formats are useful for storing GIS data that vary, such as elevation or satellite imagery. Vector formats are polygons that use points (called nodes) and lines. Vector formats are useful for storing GIS data with firm borders, such as school districts or streets.

Spatial Relationships

GIS technology can be used to display spatial relationships and linear networks. Spatial relationships may display topography, such as agricultural fields and streams. They may also display land-use patterns, such as the location of parks and housing complexes.

Linear networks, sometimes called geometric networks, are often represented by roads, rivers, and public utility grids in a GIS. A line on a map may indicate a road or highway. With GIS layers, however, that road may indicate the boundary of a school district, public park, or other demographic or land-use area. Using diverse data capture, the linear network of a river may be mapped on a GIS to indicate the stream flow of different tributaries.

GIS must make the information from all the various maps and sources align, so they fit together on the same scale. A scale is the relationship between the distance on a map and the actual distance on Earth.

Often, GIS must manipulate data because different maps have different projections. A projection is the method of transferring information from Earth's curved surface to a flat piece of paper or computer screen. Different types of projections accomplish this task in different ways, but all result in some distortion. To transfer a curved, three-dimensional shape onto a flat surface inevitably requires stretching some parts and squeezing others.

A world map can show either the correct sizes of countries or their correct shapes, but it can't do both. GIS takes data from maps that were made using different projections and combines them so all the information can be displayed using one common projection.

GIS Maps

Once all of the desired data have been entered into a GIS system, they can be combined to produce a wide variety of individual maps, depending on which data layers are included. One of the most common uses of GIS technology involves comparing natural features with human activity.

For instance, GIS maps can display what manmade features are near certain natural features, such as which homes and businesses are in areas prone to flooding.

GIS technology also allows to "dig deep" in a specific area with many kinds of information. Maps of a single city or neighborhood can relate such information as average income, book sales, or voting patterns. Any GIS data layer can be added or subtracted to the same map.

GIS maps can be used to show information about numbers and density. For example, GIS can show how many doctors there are in a neighborhood compared with the area's population.

With GIS technology, researchers can also look at change over time. They can use satellite data to study topics such as the advance and retreat of ice cover in polar regions, and how that coverage has changed through time. A police precinct might study changes in crime data to help determine where to assign officers.

One important use of time-based GIS technology involves creating time-lapse photography that shows processes occurring over large areas and long periods of time. For example, data showing the movement of fluid in ocean or air currents help scientists better understand how moisture and heat energy move around the globe.

GIS technology sometimes allows users to access further information about specific areas on a map. A person can point to a spot on a digital map to find other information stored in the GIS about that location. For example, a user might click on a school to find how many students are enrolled, how many students there are per teacher, or what sports facilities the school has.

GIS systems are often used to produce three-dimensional images. This is useful, for example, to geologists studying earthquake faults.

GIS technology makes updating maps much easier than updating maps created manually. Updated data can simply be added to the existing GIS program. A new map can then be printed or displayed on screen. This skips the traditional process of drawing a map, which can be time-consuming and expensive.

Geographic Information System and Spatial Analysis

Spatial analysis is the vital part of GIS. It can be done in two ways. One is the vector-based and the other is raster-based analysis. Since the advent of GIS in the 1980s, many government agencies have invested heavily in GIS installations, including the purchase of hardware and software and the construction of mammoth databases. Two fundamental functions of GIS have been widely realized: generation of maps and generation of tabular reports.

Indeed, GIS provides a very effective tool for generating maps and statistical reports from a database. However, GIS functionality far exceeds the purposes of mapping and report compilation. In addition to the basic functions related to automated cartography and data base management systems, the most important uses of GIS are spatial analysis capabilities. As spatial information is organized in a GIS, it should be able to answer complex questions regarding space.

Making maps alone does not justify the high cost of building a GIS. The same maps may be produced using a simpler cartographic package. Likewise, if the purpose is to generate tabular output, then a simpler database management system or a statistical package may be a more efficient solution. It is spatial analysis that requires the logical connections between attribute data and map features, and the operational procedures built on the spatial relationships among map features. These capabilities make GIS a much more powerful and cost-effective tool than automated cartographic packages, statistical packages, or data base management systems. Indeed, functions required for performing spatial analyses that are not available in either cartographic packages or data base management systems are commonly implemented in GIS.

Using GIS for Spatial Analysis

Spatial analysis in GIS involves three types of operations: Attribute Query also known as non-spatial (or spatial) query, Spatial Query and Generation of new data sets from the original database. The scope of spatial analysis ranges from a simple query about the spatial phenomenon to complicated combinations of attribute queries, spatial queries, and alterations of original data.

Attribute Query: Requires the processing of attribute data exclusive of spatial information. In other words, it's a process of selecting information by asking logical questions

Example: From a database of a city parcel map where every parcel is listed with a land use code, a simple attribute query may require the identification of all parcels for a specific land use type. Such a query can be handled through the table without referencing the parcel map. Because no spatial information is required to answer this question, the query is considered an attribute query. In this example, the entries in the attribute table that have land use codes identical to the specified type are identified.

Parcel No.	Size	Value	Land Use
102	7,500	200,000	Commercial
103	7,500	160,000	Residential
104	9,000	250,000	Commercial
105	6,600	125,000	Residential

A sample parcel map   Attribute table of the sample parcel map

Listing of Parcel No. and value with land use = 'commercial' is an attribute query. Identification of all parcels within 100-m distance is a spatial query.

Spatial Query: Involves selecting features based on location or spatial relationships, which requires processing of spatial information. For instance a question may be raised about parcels within one mile of the freeway and each parcel. In this case, the answer can be obtained either from a hardcopy map or by using a GIS with the required geographic information.

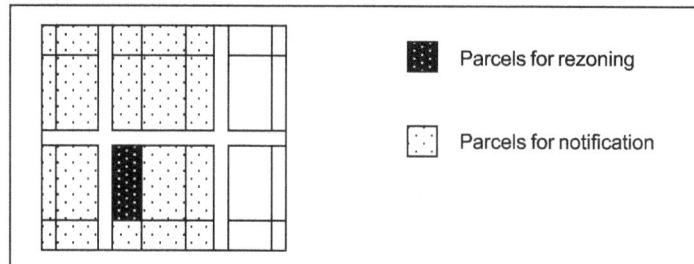

Land owners within a specified distance from the parcel to
be rezoned identified through spatial query.

Example: Let us take one spatial query example where a request is submitted for rezoning, all owners whose land is within a certain distance of all parcels that may be rezoned must be notified for public hearing. A spatial query is required to identify all parcels within the specified distance. This process cannot be accomplished without spatial information. In other words, the attribute table of the database alone does not provide sufficient information for solving problems that involve location.

While basic spatial analysis involves some attribute queries and spatial queries, complicated analysis typically require a series of GIS operations including multiple attribute and spatial queries, alteration of original data, and generation of new data sets. The methods for structuring and organizing such operations are a major concern in spatial analysis. An effective spatial analysis is one in which the best available methods are appropriately employed for different types of attribute queries, spatial queries, and data alteration.

GIS usage in Spatial Analysis

GIS can interrogate geographic features and retrieve associated attribute information, called identification. It can generate new set of maps by query and analysis. It also evolves new information by spatial operations. Here are described some analytical procedures applied with a GIS. GIS operational procedure and analytical tasks that are particularly useful for spatial analysis include:

- Single layer operations,

- Multi layer operations/Topological overlay,

- Spatial modeling,

- Geometric modeling:

  ◦ Calculating the distance between geographic features,

  ◦ Calculating area, length and perimeter,

  ◦ Geometric buffers.

- Point pattern analysis,

- Network analysis,

- Surface analysis,

- Raster/Grid analysis,

- Fuzzy Spatial Analysis,

- Geostatistical Tools for Spatial Analysis.

Single layer operations are procedures, which correspond to queries and alterations of data that operate on a single data layer.

Example: Creating a buffer zone around all streets of a road map is a single layer operation as shown in the figure.

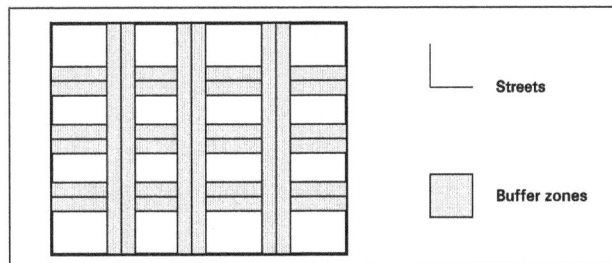

Buffer zones extended from streets.

1. Multi layer operations: Are useful for manipulation of spatial data on multiple data layers. Figure depicts the overlay of two input data layers representing soil map and a land use map respectively. The overlay of these two layers produces the new map of different combinations of soil and land use.

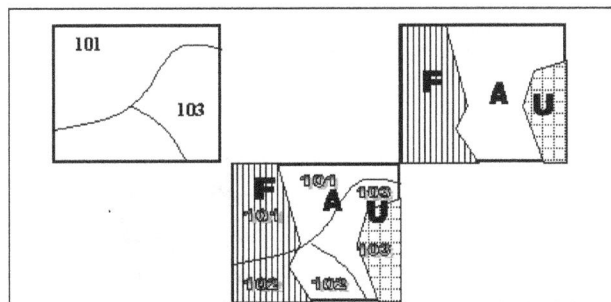

The overlay of two data layers creates a map of combined polygons.

2. Topological overlays: These are multi layer operations, which allow combining features from different layers to form a new map and give new information and features that were not present in the individual maps.

3. Point pattern analysis: It deals with the examination and evaluation of spatial patterns and the processes of point features. A typical biological survey map is shown in figure, in which each point feature denotes the observation of an endangered species such as big horn sheep in southern

California. The objective of illustrating point features is to determine the most favourable environmental conditions for this species. Consequently, the spatial distribution of species can be examined in a point pattern analysis. If the distribution illustrates a random pattern, it may be difficult to identify significant factors that influence species distribution. However, if observed locations show a systematic pattern such as the clusters in this diagram, it is possible to analyze the animals' behaviour in terms of environmental characteristics. In general, point pattern analysis is the first step in studying the spatial distribution of point features.

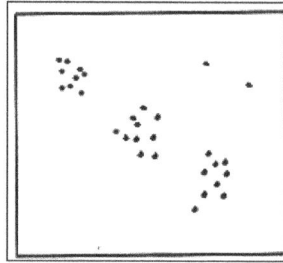

Distribution of an endangered species
examined in a point pattern analysis.

4. Network analysis: Designed specifically for line features organized in connected networks, typically applies to transportation problems and location analysis such as school bus routing, passenger plotting, walking distance, bus stop optimization, optimum path finding etc.

Figure shows a common application of GIS-based network analysis. Routing is a major concern for the transportation industry. For instance, trucking companies must determine the most cost-effective way of connecting stops for pick-up or delivery. In this example, a route is to be delineated for a truck to pick up packages at five locations. A routing application can be developed to identify the most efficient route for any set of pick-up locations. The highlighted line represents the most cost-effective way of linking the five locations.

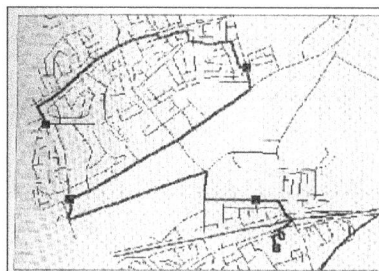

The most cost effective route links five
point locations on the street map.

Surface analysis deals with the spatial distribution of surface information in terms of a three-dimensional structure.

The distribution of any spatial phenomenon can be displayed in a threedimensional perspective diagram for visual examination. A surface may represent the distribution of a variety of phenomena, such as population, crime, market potential, and topography, among many others. The perspective diagram in figure represents topography of the terrain, generated from digital elevation model (DEM) through a series of GIS-based operations in surface analysis.

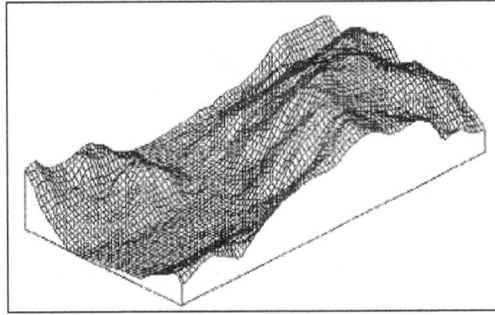

Perspective diagram representing topography
of the terrain derived from a surface analysis.

Grid analysis involves the processing of spatial data in a special, regularly spaced form. The following illustration figure shows a grid-based model of fire progression. The darkest cells in the grid represent the area where a fire is currently underway. A fire probability model, which incorporates fire behaviour in response to environmental conditions such as wind and topography, delineates areas that are most likely to burn in the next two stages. Lighter shaded cells represent these areas. Fire probability models are especially useful to fire fighting agencies for developing quick-response, effective suppression strategies.

In most cases, GIS software provides the most effective tool for performing the above tasks.

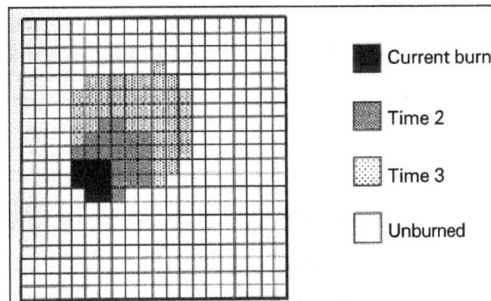

A fire behaviour model delineates areas of fire
progression based on a grid analysis.

Fuzzy Spatial Analysis

Fuzzy spatial analysis is based on Fuzzy set theory. Fuzzy set theory is a generalization of Boolean algebra to situations where zones of gradual transition are used to divide classes, instead of conventional crisp boundaries. This is more relevant in many cases where one considers 'distance to certain zone' or 'distance to road', in which case the influence of this factor is more likely to be some function of distance than a binary 'yes' or 'no'. Also in fuzzy theory maps are prepared showing gradual change in the variable from very high to very low, which is a true representation of the real world.

As stated above, the conventional crisp sets allow only binary membership function (i.e. true or false), whereas a fuzzy set is a class that admits the possibility of partial membership, so fuzzy sets are generalization of crisp sets to situations where the class membership or class boundaries are not, or cannot be, sharply defined.

Applications

Data integration using fuzzy operators using standard rules of fuzzy algebra one can combine various thematic data layers, represented by respective membership values.

Example: In a grid cell/pixel if a particular litho-unit occurs in combination with a thrust/fault, its membership value should be much higher compared with individual membership values of litho-unit or thrust/fault. This is significant as the effect is expected to be "increasive" in our present consideration and it can be calculated by fuzzy algebraic sum. Similarly, if the presence of two or a set of parameters results in "decreasive" effect, it can be calculated by fuzzy algebraic product. Besides this, fuzzy algebra offers various other methods to combine different data sets for landslide hazard zonation map preparation. To combine number of exploration data sets, five such operators exist, namely the fuzzy AND, the fuzzy OR, fuzzy algebraic product, fuzzy algebraic sum and fuzzy gamma operator.

Fuzzy logic can also be used to handle mapping errors or uncertainty, i.e. errors associated with clear demarcation of boundaries and also errors present in the area where limited ground truth exists in studies such as landslide hazard zonation. The above two kinds of errors are almost inherent to the process of data collection from different sources including remote sensing.

Geostatistical Tools for Spatial Analysis

Geostatistics studies spatial variability of regionalized variables: Variables that have an attribute value and a location in a two or three-dimensional space. Tools to characterize the spatial variability are:

- Spatial Autocorrelation Function,
- Variogram.

A variogram is calculated from the variance of pairs of points at different separation. For several distance classes or lags, all point pairs are identified which matches that separation and the variance is calculated. Repeating this process for various distance classes yields a variogram. These functions can be used to measure spatial variability of point data but also of maps or images.

Spatial Auto-correlation of Point Data

The statistical analysis referred to as spatial auto-correlation, examines the correlation of a random process with itself in space. Many variables that have discrete values measured at several specific geographic positions (i.e., individual observations can be approximated by dimensionless points) can be considered random processes and can thus be analyzed using spatial auto-correlation analysis. Examples of such phenomena are: Total amount of rainfall, toxic element concentration, grain size, elevation at triangulated points, etc.

The spatial auto-correlation function, shown in a graph is referred to as spatial auto-correlogram, showing the correlation between a series of points or a map and itself for different shifts in space or time. It visualizes the spatial variability of the phenomena under study. In general, large numbers of pairs of points that are close to each other on average have a lower variance (i.e., are better correlated), than pairs of points at larger separation. The autocorrelogram quantifies this relationship and allows gaining insight into the spatial behaviour of the phenomenon under study.

Point Interpolation

A point interpolation performs an interpolation on randomly distributed point values and returns regularly distributed point values. The various interpolation methods are: Voronoi Tesselation, moving average, trend surface and moving surface.

Example: Nearest Neighbor (Voronoi Tessellation)-In this method the value, identifier, or class name of the nearest point is assigned to the pixels. It offers a quick way to obtain a Thiessen map from point data.

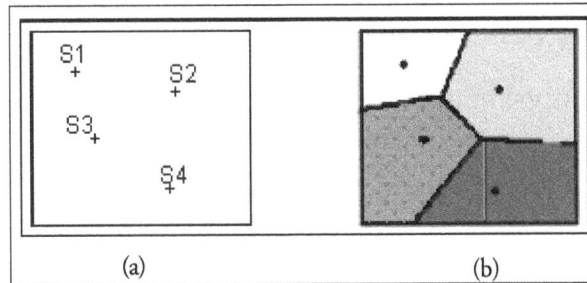

(a) An input point map, (b) The output map obtained as the result
of the interpolation operation applying the Voronoi Tessellation method.

Vector based Spatial Data Analysis

In this topic the basic concept of various vector operations are dealt in detail. There are multi layer operations, which allow combining features from different layers to form a new map and give new information and features that were not present in the individual maps.

Topological overlay:

Selective overlay of polygons, lines and points enables the users to generate a map containing features and attributes of interest, extracted from different themes or layers. Overlay operations can be performed on both raster (or grid) and vector maps. In case of raster map calculation tool is used to perform overlay. In topological overlays polygon features of one layer can be combined with point, line and polygon features of a layer.

Polygon-in-polygon overlay:

- Output is polygon coverage.
- Coverages are overlaid two at a time.
- There is no limit on the number of coverages to be combined.
- New File Attribute Table is created having information about each newly created feature.

Line-in-polygon overlay:

- Output is line coverage with additional attribute.
- No polygon boundaries are copied.
- New arc-node topology is created.

Point-in-polygon overlay:

- Output is point coverage with additional attributes.

- No new point features are created.

- No polygon boundaries are copied.

Logical Operators: Overlay analysis manipulates spatial data organized in different layers to create combined spatial features according to logical conditions specified in Boolean algebra with the help of logical and conditional operators. The logical conditions are specified with operands (data elements) and operators (relationships among data elements).

In vector overlay, arithmetic operations are performed with the help of logical operators. There is no direct way to it.

Common logical operators include AND, OR, XOR (Exclusive OR), and NOT. Each operation is characterized by specific logical checks of decision criteria to determine if a condition is true or false. Table shows the true/false conditions of the most common Boolean operations. In this table, A and B are two operands. One (1) implies a true condition and zero (0) implies false. Thus, if the A condition is true while the B condition is false, then the combined condition of A and B is false, whereas the combined condition of A OR B is true.

AND - Common Area/Intersection/Clipping Operation

OR   - Union Or Addition

NOT - (Inverter)

XOR - Minus

Truth Table of Common Boolean Operations

A	B	A AND B	A OR B	A NOT B	B NOT A	A XOR B
0	0	0	0	0	0	0
0	1	0	1	0	1	1
1	0	0	1	1	0	1
1	1	1	1	0	0	0

The most common basic multi layer operations are union, intersection, and identify operations. All three operations merge spatial features on separate data layers to create new features from the original coverage. The main difference among these operations is in the way spatial features are selected for processing.

Overlay Operations

The figure shows different types of vector overlay operations and gives flexibility for geographic data manipulation and analysis. In polygon overlay, features from two map coverages are geometrically intersected to produce a new set of information. Attributes for these new features are derived from the attributes of both the original coverages, thereby contain new spatial and attribute data relationships.

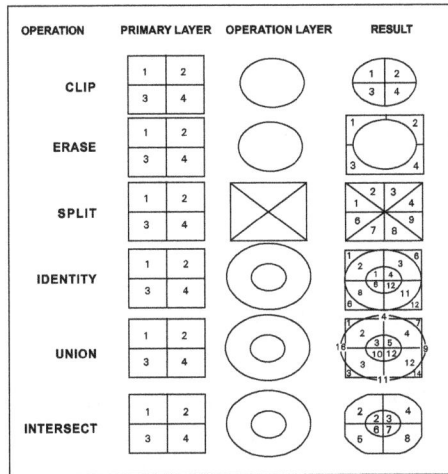

Overlay operations.

One of the overlay operation is AND (or INTERSECT) in vector layer operations, in which two coverages are combined. Only those features in the area common to both are preserved. Feature attributes from both coverages are joined in the output coverage.

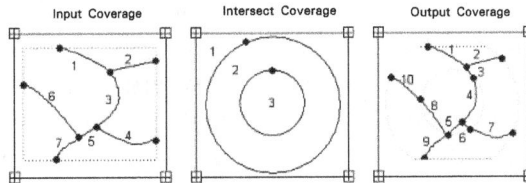

Input Coverage	
#	Attribute
1	A
2	B
3	A
4	C
5	A
6	D
7	A

Intersect Coverage	
#	Attribute
1	
2	102
3	103

Output Coverage	Input Coverage		Interest Coverage	
#	#	Attribute	#	Attribute
1	1	A	2	102
2	2	B	2	102
3	3	A	2	102
4	3	A	3	103

5	5	A	3	103
6	4	C	3	103
7	4	C	2	102
8	6	D	3	103
9	7	A	2	102
10	6	D	2	102

Raster Based Spatial Data Analysis

Present section discusses operational procedures and quantitative methods for the analysis of spatial data in raster format. In raster analysis, geographic units are regularly spaced, and the location of each unit is referenced by row and column positions. Because geographic units are of equal size and identical shape, area adjustment of geographic units is unnecessary and spatial properties of geographic entities are relatively easy to trace. All cells in a grid have a positive position reference, following the left-to-right and top-to-bottom data scan. Every cell in a grid is an individual unit and must be assigned a value. Depending on the nature of the grid, the value assigned to a cell can be an integer or a floating point. When data values are not available for particular cells, they are described as NODATA cells. NODATA cells differ from cells containing zero in the sense that zero value is considered to be data.

The regularity in the arrangement of geographic units allows for the underlying spatial relationships to be efficiently formulated. For instance, the distance between orthogonal neighbors (neighbors on the same row or column) is always a constant whereas the distance between two diagonal units can also be computed as a function of that constant. Therefore, the distance between any pair of units can be computed from differences in row and column positions. Furthermore, directional information is readily available for any pair of origin and destination cells as long as their positions in the grid are known.

Advantages of using the Raster Format in Spatial Analysis

Efficient processing: Because geographic units are regularly spaced with identical spatial properties, multiple layer operations can be processed very efficiently.

Numerous existing sources: Grids are the common format for numerous sources of spatial information including satellite imagery, scanned aerial photos, and digital elevation models, among others. These data sources have been adopted in many GIS projects and have become the most common sources of major geographic databases.

Different feature types organized in the same layer: For instance, the same grid may consist of point features, line features, and area features, as long as different features are assigned different values.

Grid Format Disadvantages

- Data redundancy: When data elements are organized in a regularly spaced system, there is a data point at the location of every grid cell, regardless of whether the data element is needed or not. Although, several compression techniques are available, the advantages of gridded data are lost whenever the gridded data format is altered through compression. In most cases, the compressed data cannot be directly processed for analysis. Instead, the compressed raster data must first be decompressed in order to take advantage of spatial regularity.

- Resolution confusion: Gridded data give an unnatural look and unrealistic presentation unless the resolution is sufficiently high. Conversely, spatial resolution dictates spatial properties. For instance, some spatial statistics derived from a distribution may be different, if spatial resolution varies, which is the result of the well-known scale problem.

- Cell value assignment difficulties: Different methods of cell value assignment may result in quite different spatial patterns.

Participatory Geographic Information System

Participatory GIS (PGIS) is a participatory approach to spatial planning and spatial information and communications management.

PGIS combines Participatory Learning and Action (PLA) methods with geographic information systems (GIS). PGIS combines a range of geo-spatial information management tools and methods such as sketch maps, participatory 3D modelling (P3DM), aerial photography, satellite imagery, and global positioning system (GPS) data to represent peoples' spatial knowledge in the forms of (virtual or physical) two- or three-dimensional maps used as interactive vehicles for spatial learning, discussion, information exchange, analysis, decision making and advocacy. Participatory GIS implies making geographic technologies available to disadvantaged groups in society in order to enhance their capacity in generating, managing, analysing and communicating spatial information.

PGIS practice is geared towards community empowerment through measured, demand-driven, user-friendly and integrated applications of geo-spatial technologies. GIS-based maps and spatial analysis become major conduits in the process. A good PGIS practice is embedded into long-lasting spatial decision-making processes, is flexible, adapts to different socio-cultural and bio-physical environments, depends on multidisciplinary facilitation and skills and builds essentially on visual language. The practice integrates several tools and methods whilst often relying on the combination of 'expert' skills with socially differentiated local knowledge. It promotes interactive participation of stakeholders in generating and managing spatial information and it uses information about specific landscapes to facilitate broadly-based decision making processes that support effective communication and community advocacy.

If appropriately utilized, the practice could exert profound impacts on community empowerment, innovation and social change. More importantly, by placing control of access and use of culturally sensitive spatial information in the hands of those who generated them, PGIS practice could protect traditional knowledge and wisdom from external exploitation.

CyberGIS

CyberGIS, or cyber geographic information science and systems, is an interdisciplinary field combining cyberinfrastructure, e-science, and geographic information science and systems (GIS). Cyber-GIS has a particular focus on computational and data-intensive geospatial problem-solving within various research and education domains. The need for GIS has extended beyond traditional forms of geographic analysis and study, which includes adapting to new sources and kinds of data, high-performance computing resources, and online platforms based on existing and emerging information networks. The name cyberGIS first appeared in Geographic Information Science literature in 2010. CyberGIS is characterized as digital geospatial ecosystems. These systems are developed and have evolved through heterogeneous computing environments, as well as human communication and information environments. CyberGIS can be considered a new generation of geographic information systems (GIS). These systems are based on advanced computing and information infrastructure, which analyze and model geospatial data, providing computationally intensive spatial analysis, modeling, and collaborative geospatial problem-solving at previously unprecedented scales.

Further descriptions of CyberGIS include: "a fundamentally new software framework comprising a seamless integration of cyberinfrastructure, GIS, and spatial analysis/modeling capabilities", and a "GIS detached from the desktop and deployed on the web, with the associated issues of hardware, software, data storage, digital networks, people, training and education." Earlier scientific research demonstrated problems for integration of spatiotemporal data and analytics due to geographic and spatial complexity. CyberGIS attempts to move beyond traditional scientific and technical constraints within conventional GIS by innovating computer and data-intensive cyber environments, exploiting spatiotemporal characteristics inherent in various scientific domains, and using big data and high-performance computing approaches to collaborative problem solving.

In 2004, early research on middleware that integrates GIS with high performance and distributed computing technologies laid the foundation for the subsequent research and development of cyberGIS. This middleware was called geo-middleware because it was tailored to solving geographic problems. Geo-middleware was aimed at enabling collaborative problem-solving and decision-making by taking advantage of massive computational resources provided by high-performance computing infrastructure. Data-intensive spatiotemporal analytics and simulation require substantial computing power while geographic analysis often needs to address the effects of scale and spatial relationships have on various complex phenomena. This means that cyberGIS is focused on cutting-edge GIS advances that are dependent on advanced cyberinfrastructure and high-performance computing instead of conventional sequential computing and GIS approaches. By exploiting massive cyber resources, it is possible for cyberGIS to resolve broader scientific challenges through data-intensive spatiotemporal knowledge discovery.

The scientific cyberinfrastructure, geospatial, and GIS communities have been working extensively to advance the field of cyberGIS. A key community event was the National Science Foundation TeraGrid Workshop on cyberGIS that took place in conjunction with the University Consortium for Geographic Information Science Winter Meeting of February 2010. This workshop report laid out a compelling cyberGIS roadmap that articulates fundamental issues of cyberGIS for innovating cyberinfrastructure; at the same time, the report contributes to the advancement of the next-generation GIS, which integrates high performance computing, distributed computing, and Internet-enabled collaborative capabilities for geospatial discovery and innovation.

In 2010, Dr. Shaowen Wang described the first cyberGIS framework for the synthesis of cyberinfrastructure, GIS, and spatial analysis. Since then, it has been recognized as an important area for advanced cyberinfrastructure and GIS research. A multi-institution and multidisciplinary initiative, funded by the National Science Foundation in 2010, began a six-year $4.8 million project on

"CyberGIS Software Integration for Sustained Geospatial Innovation." This major initiative has established three interrelated pillars of a cutting-edge cyberGIS software environment: the Cyber-GIS Gateway, CyberGIS Toolkit, and GISolve middleware.

The ROGER supercomputer.

CyberGIS Supercomputer

In 2014, the CyberGIS Center for Advanced Digital and Spatial Studies at the University of Illinois at Urbana-Champaign received a National Science Foundation major research instrumentation grant to establish ROGER as the first cyberGIS supercomputer. ROGER, hosted by the National Center for Supercomputing Applications, is optimized to deal with geospatial data and computation and is equipped with:

- Approximately six petabytes of raw disk storage with high input/output bandwidth;

- Solid-state drives for applications demanding high data-access performance;

- Advanced graphics processing units for exploiting massive parallelism in geospatial computing;

- Interactive visualization supported with a high-speed network and dynamically provisioned cloud computing resources.

CyberGIS software and tools integrate these system components to support a large number of users who are investigating scientific problems in areas spanning biosciences, engineering, geosciences, and social sciences.

Applications and Uses

GIS in Mapping: Mapping is a central function of Geographic Information System, which provides a visual interpretation of data. GIS store data in database and then represent it visually in a mapped format. People from different professions use map to communicate. It is not necessary to be a skilled cartographer to create maps. Google map, Bing map, Yahoo map are the best example for web based GIS mapping solution.

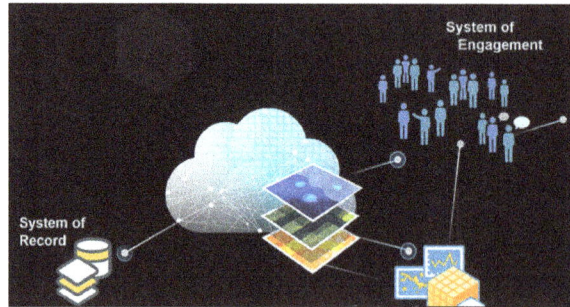

Telecom and Network services: GIS can be a great planning and decision making tool for telecom industries. GDi GISDATA enables wireless telecommunication organizations to incorporate geographic data in to the complex network design, planning, optimization, maintenance and activities. This technology allows telecom to enhance a variety of application like engineering application, customer relationship management and location based services.

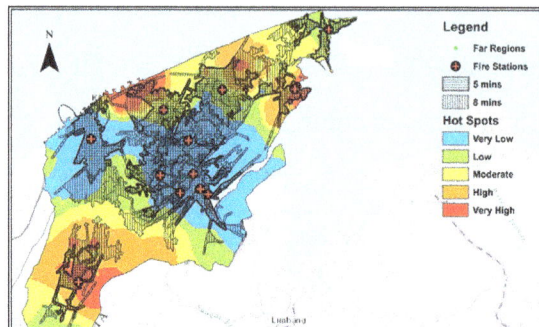

Accident Analysis and Hot Spot Analysis: GIS can be used as a key tool to minimize accident hazard on roads, the existing road network has to be optimized and also the road safety measures have to be improved. This can be achieved by proper traffic management. By identifying the accident locations, remedial measures can be planned by the district administrations to minimize the accidents in different parts of the world. Rerouting design is also very convenient using GIS.

Urban Planning: GIS technology is used to analyze the urban growth and its direction of expansion, and to find suitable sites for further urban development. In order to identify the sites suitable for the urban growth, certain factors have to consider which is: land should have proper accessibility, land should be more or less flat, land should be vacant or having low usage value presently and it should have good supply of water.

Transportation Planning: GIS can be used in managing transportation and logistical problems. If transport department is planning for a new railway or a road route then this can be performed by adding environmental and topographical data into the GIS platform. This will easily output the best route for the transportation based on the criteria like flattest route, least damage to habitats and least disturbance from local people. GIS can also help in monitoring rail systems and road conditions.

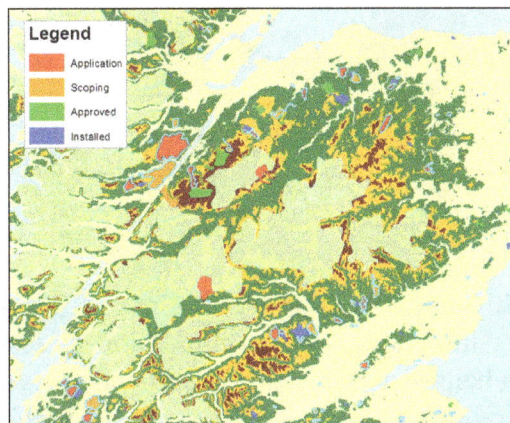

Environmental Impact Analysis: EIA is an important policy initiative to conserve natural resources and environment. Many human activities produce potential adverse environmental effects which include the construction and operation of highways, rail roads, pipelines, airports, radioactive waste disposal and more. Environmental impact statements are usually required to contain specific information on the magnitude and characteristics of environmental impact. The EIA can be carried out efficiently by the help of GIS, by integrating various GIS layers, assessment of natural features can be performed.

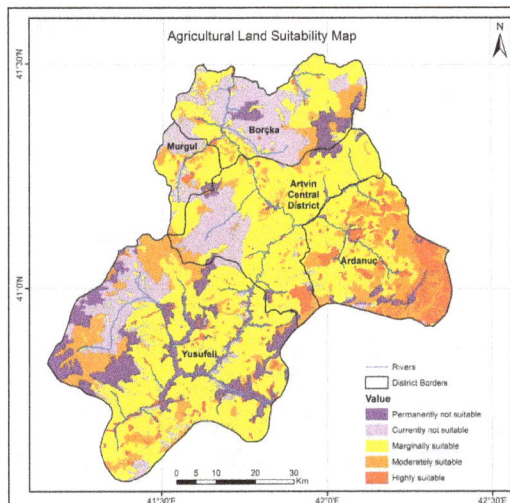

Agricultural Applications: GIS can be used to create more effective and efficient farming techniques. It can also analyze soil data and to determine: what are the best crop to plant?, where they should go? how to maintain nutrition levels to best benefit crop to plant?. It is fully integrated and widely accepted for helping government agencies to manage programs that support farmers and protect the environment. This could increase food production in different parts of the world so the world food crisis could be avoided.

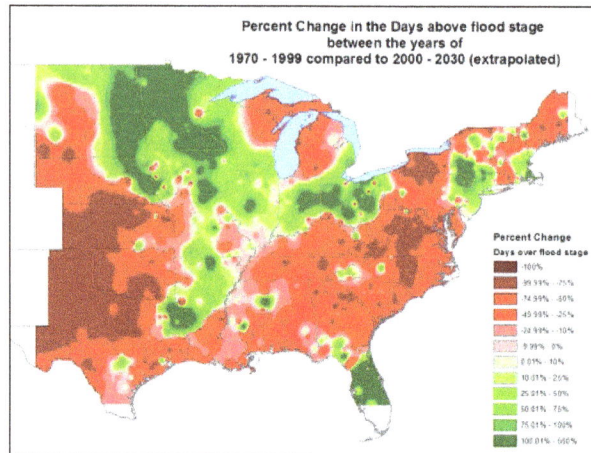

Disaster Management and Mitigation: Today a well-developed GIS systems are used to protect the environment. It has become an integrated, well developed and successful tool in disaster management and mitigation. GIS can help with risk management and analysis by displaying which areas are likely to be prone to natural or man-made disasters. When such disasters are identified, preventive measures can be developed.

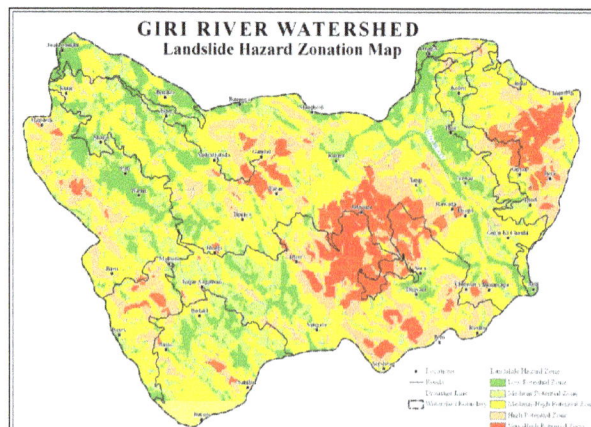

Landslide Hazard Zonation using GIS: Landslide hazard zonation is the process of ranking different parts of an area according to the degrees of actual or potential hazard from landslides. The evaluation of landslide hazard is a complex task. It has become possible to efficiently collect, manipulate and integrate a variety of spatial data such as geological, structural, surface cover and slope characteristics of an area, which can be used for hazard zonation. The entire above said layer can well integrate using GIS and weighted analysis is also helpful to find Landslide prone area. By the help of GIS we can do risk assessment and can reduce the losses of life and property.

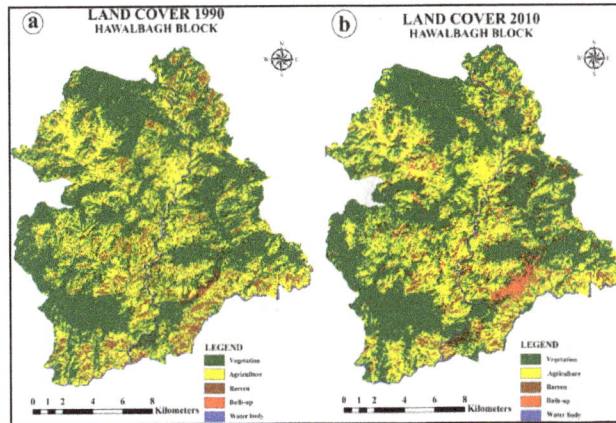

Determine Land Use/Land Cover Changes: Land cover means the feature that is covering the barren surface. Land use means the area in the surface utilized for particular use. The role of GIS technology in land use and land cover applications is that we can determine land use/land cover changes in the different areas. Also it can detect and estimate the changes in the land use/land cover pattern within time. It enables to find out sudden changes in land use and land cover either by natural forces or by other activities like deforestation.

Navigation (Routing and Scheduling): Web-based navigation maps encourage safe navigation in waterway. Ferry paths and shipping routes are identified for the better routing. ArcGIS supports safe navigation system and provides accurate topographic and hydrographic data. Recently DNR's Coastal Resources Division began the task of locating, documenting, and cataloging these no historic wrecks with GIS. This division is providing public information that make citizens awareness of these vessel locations through web map. The web map will be regularly updated to keep the boating public informed of these coastal hazards to minimize risk of collision and injury.

Flood Damage Estimation: GIS helps to document the need for federal disaster relief funds, when appropriate and can be utilized by insurance agencies to assist in assessing monetary value of property loss. A local government need to map flooding risk areas for evaluate the flood potential level in the surrounding area. The damage can be well estimate and can be shown using digital maps.

Natural Resources Management: By the help of GIS technology the agricultural, water and forest resources can be well maintain and manage. Foresters can easily monitor forest condition.

Agricultural land includes managing crop yield, monitoring crop rotation, and more. Water is one of the most essential constituents of the environment. GIS is used to analyze geographic distribution of water resources. They are interrelated, i.e. forest cover reduces the storm water runoff and tree canopy stores approximately 215,000 tons carbon. GIS is also used in afforestation.

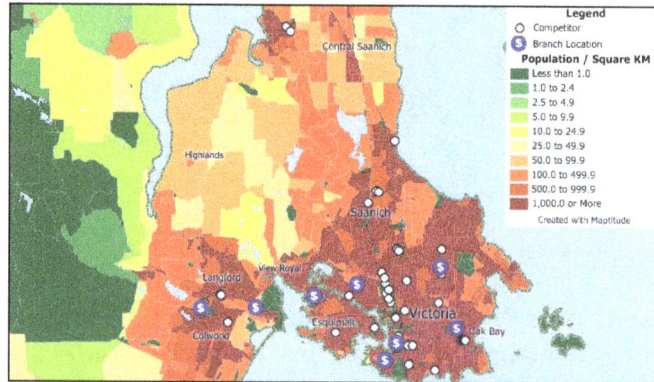

GIS Solutions in Banking Sector: Today rapid development occurs in the banking sector. So it has become more market driven and market responsive. The success of this sector largely depends on the ability of a bank to provide customer and market driven services. GIS plays an important role providing planning, organizing and decision making.

Soil Mapping: Soil mapping provides resource information about an area. It helps in understanding soil suitability for various land use activities. It is essential for preventing environmental deterioration associated with misuse of land. GIS Helps to identify soil types in an area and to delineate soil boundaries. It is used for the identification and classification of soil. Soil map is widely used by the farmers in developed countries to retain soil nutrients and earn maximum yield.

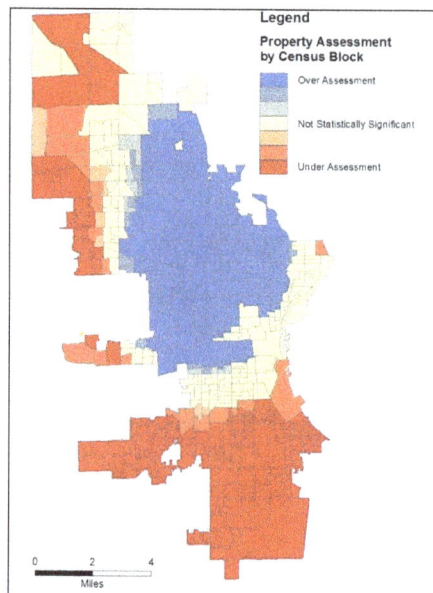

GIS Based Digital Taxation: In Local Governments, GIS is used to solve taxation problems. It is used to maximize the government income. For example, for engineering, building permits, city development and other municipal needs, GIS is used. Often the data collected and used by one agency or department can be used by another. Example Orhitec ltd can supply you with a system to manage property tax on a geographic basis that can work interactively with the municipal tax collection department. Using GIS we can develop a digital taxation system.

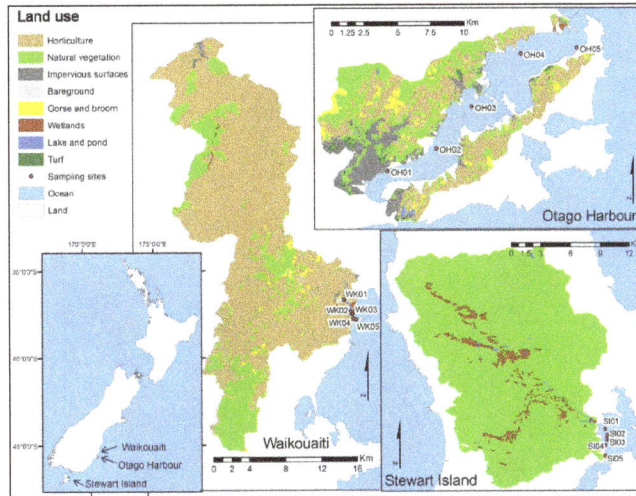

Land Information System: GIS based land acquisition management system will provide complete information about the land. Land acquisition managements is being used for the past 3 or 4 years only. It would help in assessment, payments for private land with owner details, tracking of land allotments and possessions identification and timely resolution of land acquisition related issues.

Surveying: Surveying is the measurement of location of objects on the earth's surfaces. Land survey is measuring the distance and angles between different points on the earth surface. An increasing number of national and governments and regional organizations are using GNSS measurements. GNSS is used for topographic surveys where a centimeter level accuracy is provided. These data can be incorporated in the GIS system. GIS tools can be used to estimate area and also, digital maps can prepared.

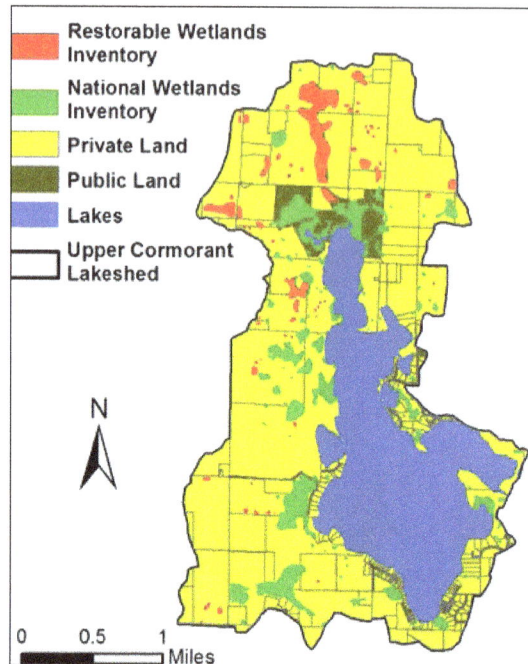

Wetland Mapping: Wetlands contribute to a healthy environment and retain water during dry periods, thus keeping the water table high and relatively stable. During the flooding they act to reduce flood levels and to trap suspended solids and attached nutrients. GIS provide options for wetland mapping and design projects for wetland conservation quickly with the help of GIS. Integration with Remote Sensing data helps to complete wetland mapping on various scale. We can create a wetland digital data bank with spices information using GIS.

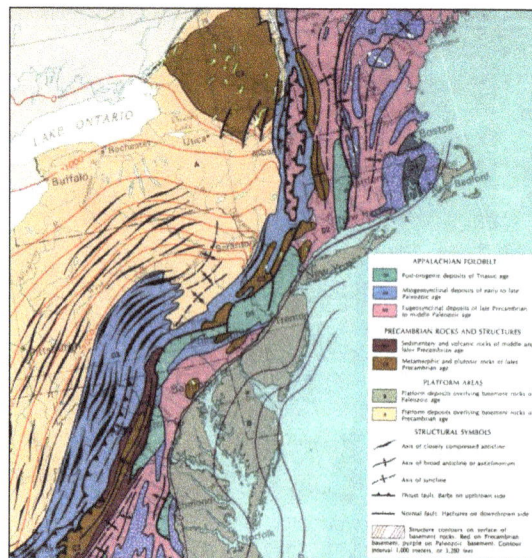

GIS Applications in Geology: Geologists use GIS in a various applications. The GIS is used to study geologic features, analyze soils and strata, assess seismic information, and or create three dimensional (3D) displays of geographic features. GIS can be also used to analyze rock information characteristics and identifying the best dam site location.

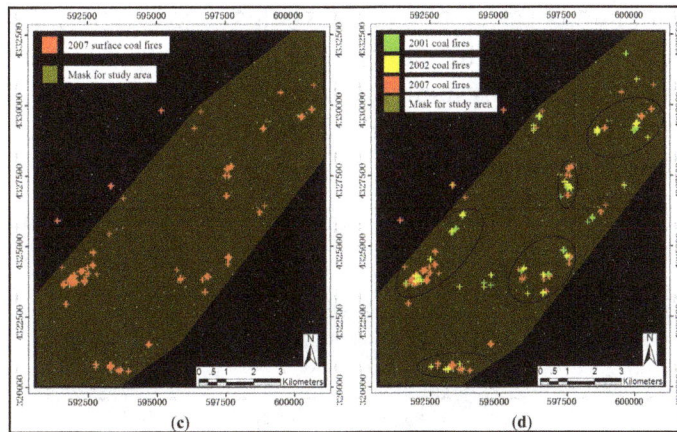

(c)   (d)

Detection of Coal Mine Fires: GIS technology is applied in the area of safe production of coal mine. Coal mine have developed an information management system, the administrators can monitor the safe production of coal mine and at the same time improve the abilities to make decisions. Fire happens frequently in coal mines. So it can assessed spontaneous combustion risk using GIS tools.

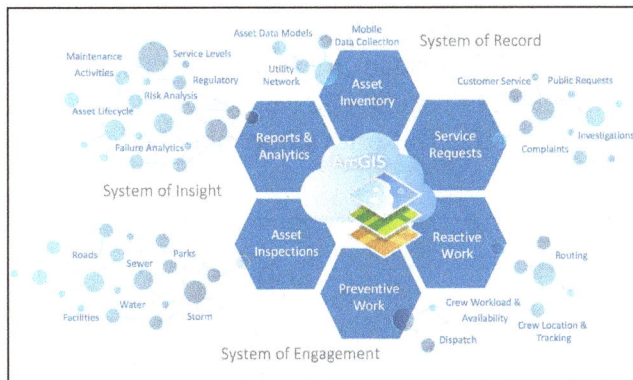

Assets Management and Maintenance: GIS helps organizations to gain efficiency even in the face of finite resources and the need to hold down the cost. Knowing the population at risk enables planners to determine where to allocate and locate resources more effectively. Operations and maintenance staff can deploy enterprise and mobile workforce. GIS build mobile applications that provide timely information in the field faster and more accurate work order processing.

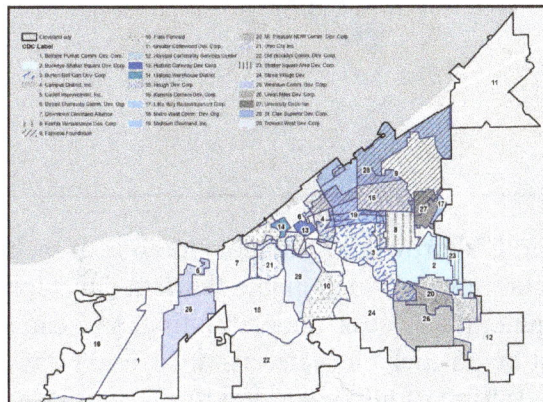

GIS for Planning and Community Development: GIS helps us to better understand our world so we can meet global challenges. Today GIS technology is advancing rapidly, providing many new capabilities and innovations in planning. By applying known part of science and GIS to solve unknown part, that helps to enhance the quality of life and achieve a better future. Creating and applying GIS tools and knowledge allow us integrating geographic intelligence into how we think and behave.

GIS in Dairy Industry: Geographic Information System is used in a various application in the dairy industry, such as distribution of products, production rate, location of shops and their selling rate. These can be monitored by using GIS system. It can be also possible to understand the demand of milk and milk products in different region. GIS can prove to be effective tool for planning and decision making for any dairy industry. These advantages has added new vistas in the field of dairy farm and management.

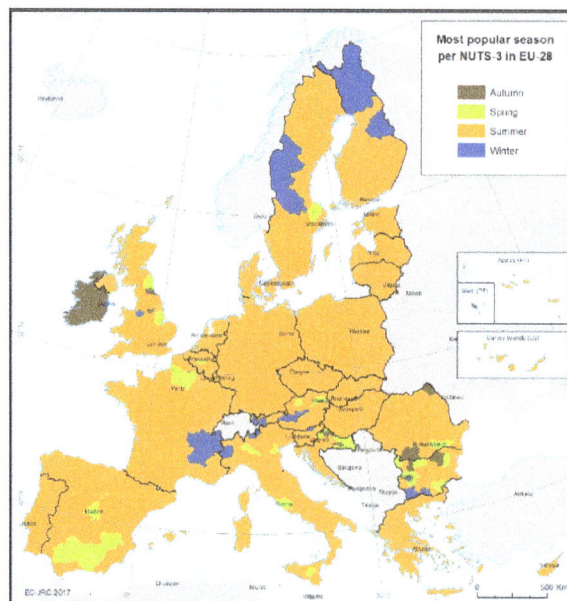

Tourism Information System: GIS provides a valuable toolbox of techniques and technologies of wide applicability to the achievement of sustainable tourism development. This provide an ideal platform tools required to generate a better understanding, and can serve the needs of tourists. They will get all the information on click, measure distance, find hotels, restaurant and even navigate to their respective links. Information plays a vital role to tourists in planning their travel from one place to another, and success of tourism industry. This can bring many advantages for both tourist and tourism department.

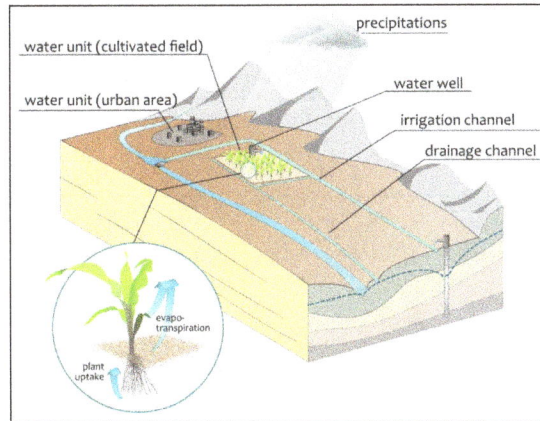

Irrigation Water Management: Water availability for irrigation purposes for any area is vital for crop production in that region. It needs to be properly and efficiently managed for the proper utilization of water. To evaluate the irrigation performance, integrated use of satellite remote sensing and GIS assisted by ground information has been found to be efficient technique in spatial and time domain for identification of major crops and their conditions, and determination of their areal extent and yield. Irrigation requirements of crop were determined by considering the factors such as evapotranspiration, Net Irrigation Requirement, Field irrigation Requirement, Gross Irrigation Requirement, and month total volume of water required, by organizing them in GIS environment.

Fire Equipment Response Distance Analysis: GIS can be used to evaluate how far (as measured as via the street network) each portion of the street network is from a firehouse. This can be useful in evaluating the best location for a new firehouse or in determining how well the fire services cover particular areas for insurance ratings.

Worldwide Earthquake Information System: One of the most frightening and destructive phenomena of nature is the occurrence of an earthquake. There is a need to have knowledge regarding the trends in earthquake occurrence worldwide. A GIS based user interface system for querying on earthquake catalogue will be of great help to the earthquake engineers and seismologists in understanding the behavior pattern of earthquake in spatial and temporal domain.

Volcanic Hazard Identification: Volcanic hazard to human life and environment include hot avalanches, hot particles gas clouds, lava flows and flooding. Potential volcanic hazard zone can be recognized by the characteristic historical records of volcanic activities, it can incorporate with GIS. Thus an impact assessment study on volcanic hazards deals with economic loss and loss of lives and property in densely populated areas. The GIS based platforms enables us to find out the damage and rapid response against volcanic activities may helps to reduce the effect in terms of wealth and health of people.

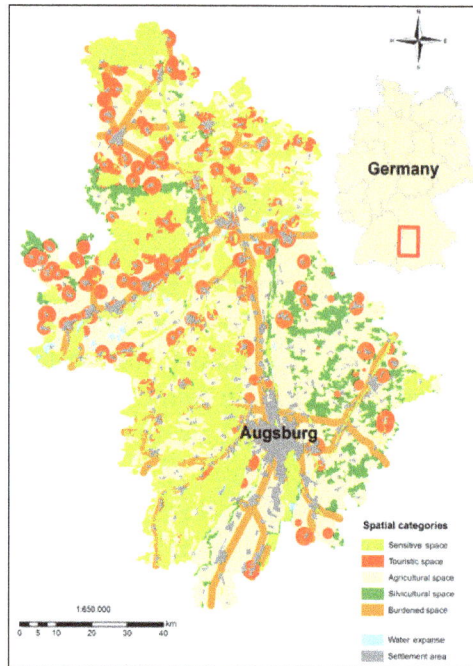

Energy Use Tracking and Planning: GIS is a valuable tool that helps in the planning organizing and subsequent growth in the energy and utilities industries. The effective management of energy systems is a complex challenge. GIS has enormous potential for planning, design and maintenance of facility. Also it provide improved services and that too cost effectively.

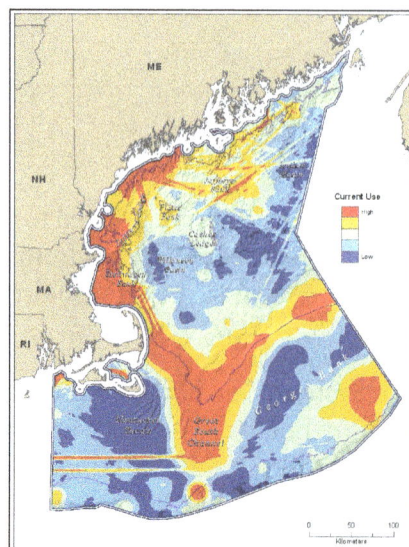

GIS for Fisheries and Ocean Industries: GIS tools add value and the capability to ocean data. Arc-Gis is used to determine the spatial data for a fisheries assessment and management system. It is extensively used in the ocean industry area and we get accurate information regarding various commercial activities. To enhance minimizing cost for the fishing industry. Also it can determine the location of illegal fishing operations.

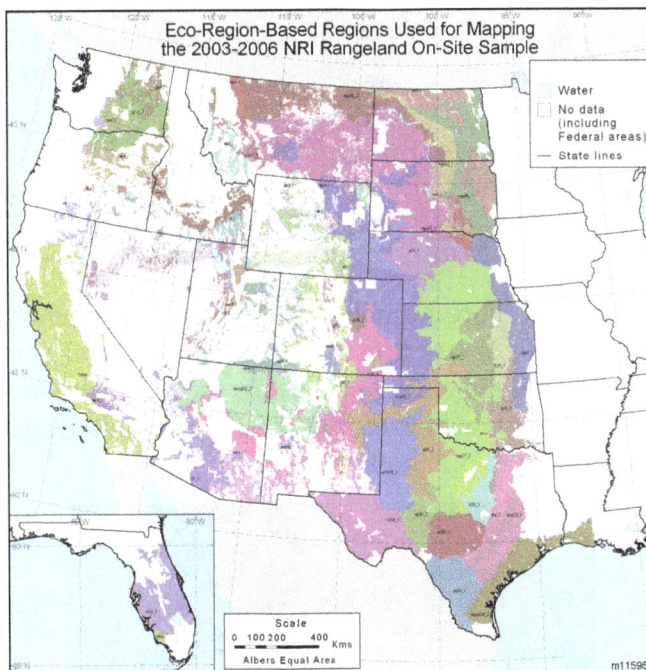

Eco-Region-Based Regions Used for Mapping the 2003-2006 NRI Rangeland On-Site Sample

Monitor Rangeland Resources: GIS is a valuable tool used to monitor the changes of rangeland resource and for evaluating its impact on environment, livestock and wild life. Accurate observation and measurements are to be made to find out the changes in the rangeland conditions. GIS is also used to monitoring ecological and seasonal rangeland conditions.

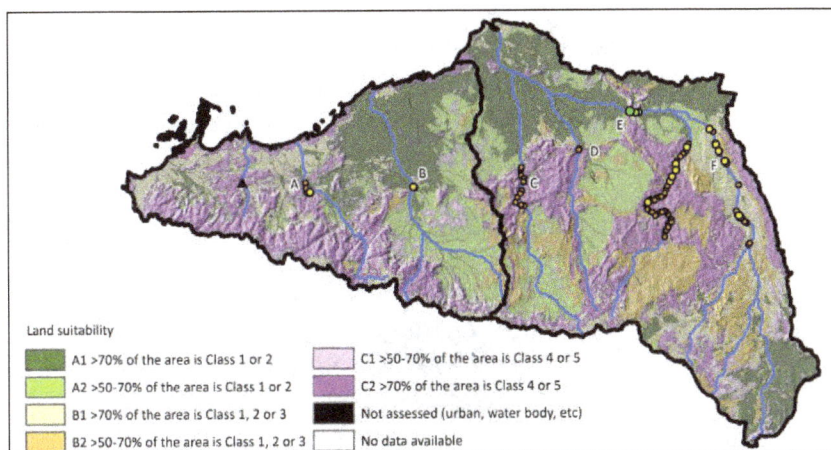

Land suitability

Reservoir Site Selection: GIS is used to find a suitable site for the dam. GIS tries to find best location that respect to natural hazards like earthquake and volcanic eruption. For the finding of dam site selection the factors include economic factors, social considerations, engineering factors and environmental problems. This all information are layered in the GIS.

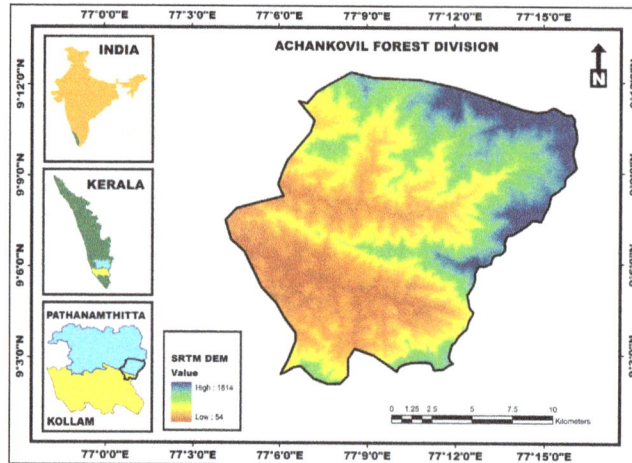

Forest Fire Hazard Zone Mapping: Forest is one of the important element of the nature. It plays an important role in the local climate. Forest fires caused extensive damage to our communities and environmental resource base. GIS can effectively use for the forest fire hazard zone mapping and also for the loss estimation. GIS also help to capture real time monitoring of fire prone areas. This is achieved by the help of GNSS and satellite Remote Sensing.

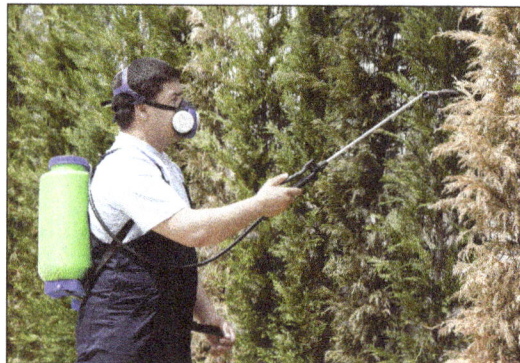

Pest Control and Management: Pest control helps in the agricultural production. Increasing in the rate of pest and weeds can lead to decrease in the crop production. Therefore GIS plays an important role to map out infested areas. This leads in the development of weed and pest management plan.

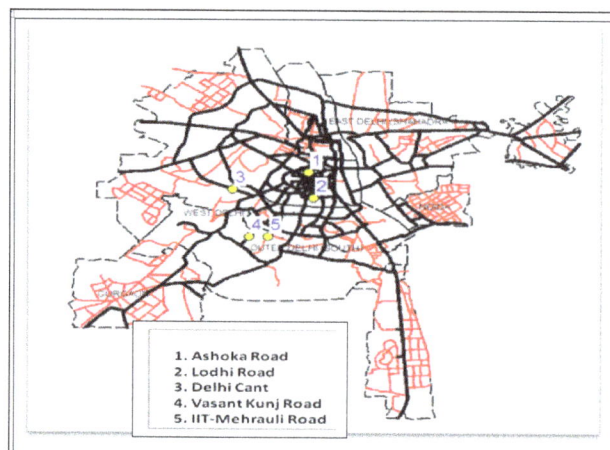

Traffic Density Studies: GIS can effectively use for the management of traffic problems. Today's population along with the road traffic is increasing exponentially. The advantage of GIS make it an attractive option to be used to face the emerging traffic problems. By creating an extensive database that has all the traffic information such as speed data, road geometry, traffic flow and other spatial data and processing this information will provide us the graphical bigger picture for the traffic management.

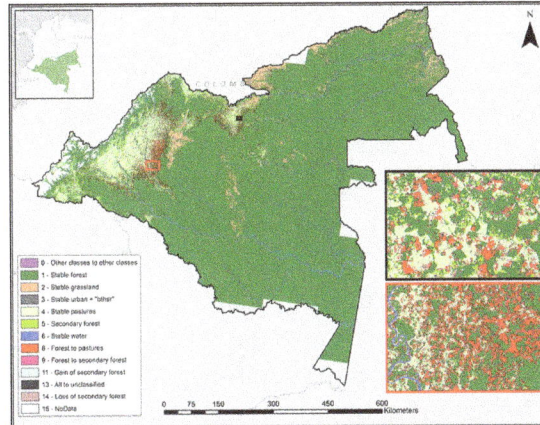

Deforestation: Nowadays forest area is decreasing every year, due to different activities. GIS is used to indicate the degree of deforestation and vital causes for the deforestation process. GIS is used to monitor deforestation.

Space Utilization: GIS helps managers to organize and spatially visualize space and how it can best be used. Operational costs can be decreased by more efficiently using space including managing the moves of personal and assets as well as the storage materials. The 3D visualization in GIS platforms helps planers to create a feeling of experience like virtual walk inside the building and rooms before construction.

Desertification: Desertification is the land degradation due to climatic variations or human activities. GIS can provide the information of degraded land which can be managed by governmental agencies or by the communities themselves. GIS plays a vital role to reduce the desertification, the local governments are now widely depends on GIS for reducing desertification. With location based GIS analysis we can find where or which area is suitable for planting new vegetation and which area for the pipeline construction.

Disaster and Business Continuity Planning: Viewing building and locations assets along with emergency information such as weather patterns, and disaster zones, can provide organizations the required information to make better decision. GIS provide holistic understanding of facility status and performance, and brings together department, business systems, and data source for a comprehensive view into and throughout the organization.

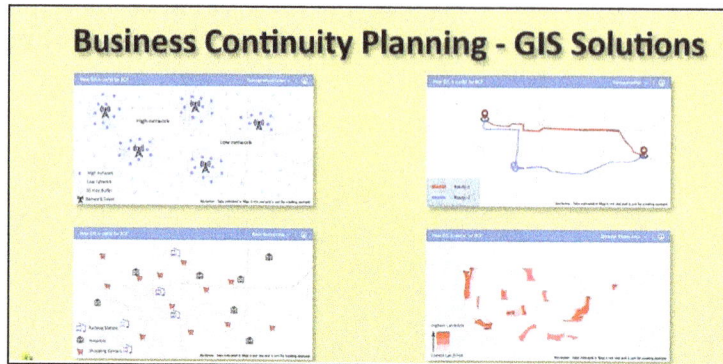

GIS for Business: GIS is also used for managing business information based on its location. GIS can keep a track of where the customers are located, site business, target marketing campaigns, and optimize sales territories and model retail spending patterns. Such an added advantage is provided by the GIS to enhance in making companies more competitive and successful.

Utilities: The GIS is used for different type of utilities like electricity, telecom and cooking gas on a daily basis and utilities to help them in mapping, in inventory systems, track maintenance, monitor regulatory compliance or model distribution analysis, transformer analysis and load analysis.

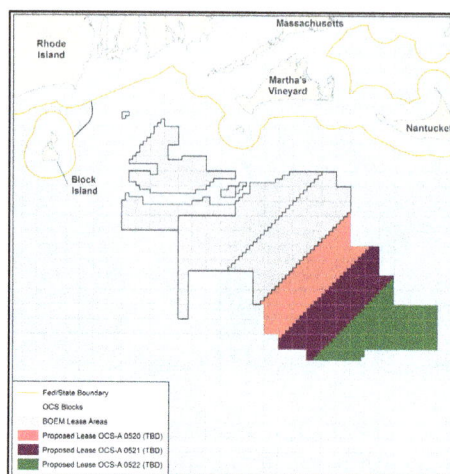

Lease Property and Management: Revenue can be increased, operations and maintenance cost can be reduced when GIS is used to help manage space. Real estate and property managers can see and make queries about space including its availability, size and special constraints for the most cost effective use.

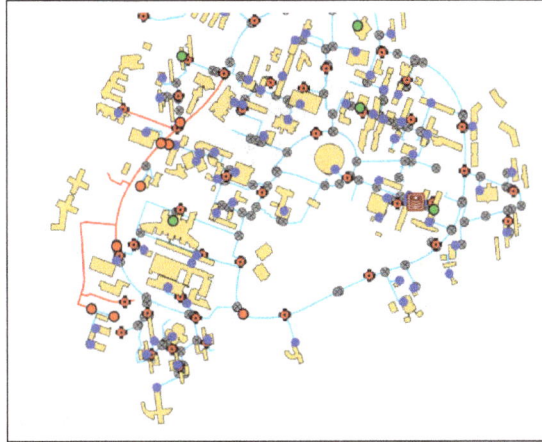

Development of Public Infrastructure Facilities: GIS has many uses and advantages in the field of facility management. GIS can be used by facility managers for space management, visualization and planning, emergency and disaster planning and response. It can be used throughout the life cycle of a facility from deciding where to build to space planning. Also it provides facilitate better planning and analysis.

GIS for Drainage Problems in Tea Plantation Areas: Drainage problem in tea plantation differ widely because of its varied nature of physical conditions. Tea crop requires moisture at adequate levels all times of its growth. Any variation either excess or lack has a direct impact on the tea yield. This become greatly influenced the productivity of tea. Required some hydraulic design to solve this problem such as design of drains, checking the adequacy of the river and classification of water logged areas etc. GIS helps to reduce the water logging by establishing well developed plans.

Collection of Information about Geographic Features: GIS is not simply a computer system used for making maps. A map is simply the most common way of reporting information from a GIS database. So these systems are not only for creating maps but also most importantly the collection of information about the geographic features such as building, roads, pipes, streams, ponds and many more that are located in your community.

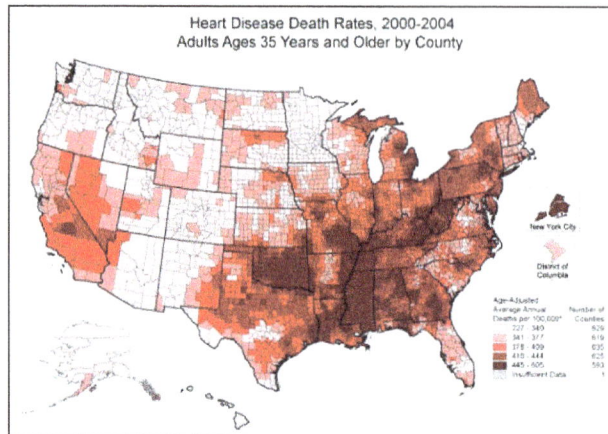

GIS for Public Health: GIS provides the cost effective tool for evaluating interventions and policies potentially affecting health outcomes. GIS analysis, environmental health data is also helpful in explaining disease patterns of relationships with social, institutional, technological and natural environment. It can be understand the complex spatial temporal relationship between environmental pollution and disease, and identifying exposures to environmental hazards. GIS can significantly add value to environmental and public health data.

Location Identification: This technique is used to find a location for a new retail outlet. It helps to find out what exists at a particular location. A location can be described in many ways, using, for instance, name of place, post code, or geographic reference such as longitude or latitude or X/Y.

Knowledge Based System for Defense Purpose: Regular analysis of terrain is essential for today's fast paced battlefield. Conventional method of studying paper topographical maps is being replaced by use of maps in digital form to get terrain information. It is increasingly being used to derive terrain information from digital images. Which help to the selection of suitable sites for various military uses more accurate and faster. The uses of GIS provide information regarding the terrain features which can be useful for planning today's war strategies.

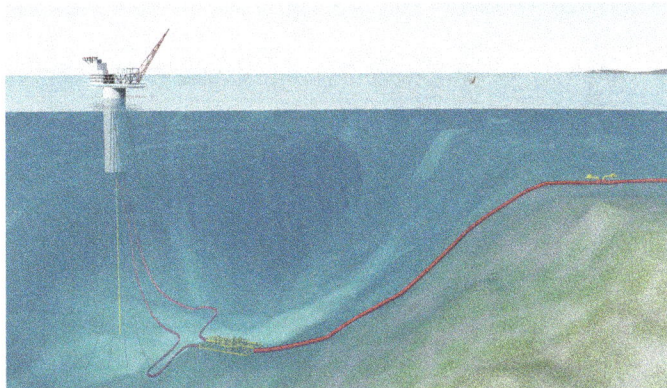

Pipeline Route Selection: Pipeline route planning and selection is usually a complex task. GIS technology is faster, better and more efficient in this complex task. Accurate pipeline route selection brings about risk and cost reduction as well as better decision making process. GIS least cost path analysis have been effectively used to determine suitable oil and gas pipeline routes. An optimal route will minimize reduce economic loss and negative socio-environmental impacts.

Producing Mailing Labels for abutter Notification: Zoning board of appeals hearings or proposed action by a town or city require notifying abutting property owners. A GIS application for producing abutter mailing labels enables you to identify abutting property owners are in different ways. Once the properties are identified this kind of GIS applications can produce mailing labels and be integrated with a word processing "mail merge" capabilities.

Site Suitability for Waste Treatment Plant: There is an increasing amount of waste due to the over population growth. This has negative impact on the environment. With the help of GIS we can integrate various aspect layers in GIS and can identify which place is suitable for waste treatment plant. This process will reduce the time and it is cost effective. Also it enhances the accuracy. It provides a GIS analyst to identify a list of suitable dumping sites for further investigations. It also provides a digital bank for future monitoring program of the site.

Geologic Mapping: GIS is an effective tool in geological mapping. It becomes easy for surveyors to create 3D maps of any area with precise and desired scaling. The results provide accurate measurements, which helps in several field where geological map is required. This is cost effective and offers more accurate data, there by easing the scaling process when studying geologic mapping.

Environment: The GIS is used every day to help protect the environment. The environmental professional uses GIS to produce maps, inventory species, measure environmental impact, or trace pollutants. The environmental applications for GIS are almost endless. It can be used to monitor the environment and analyze changes.

Infrastructure Development: Advancement and availability of technology has set a new mark for professionals in the infrastructure development area. Now more and more professionals are seeking help of these technologically smart and improved information systems like GIS for infrastructure development. Each and every phase of infrastructure life cycle is greatly affected and enhanced by the enrollment of GIS.

Coastal Development and Management: The coastal zone represents varied and highly productive ecosystem such as mangrove, coral reefs, see grasses and sand dunes. GIS could be generating data required for macro and micro level planning of coastal zone management. GIS could be used in creating baseline inventory of mapping and monitoring coastal resources, selecting sites for brackish water aquaculture, studying coastal land forms.

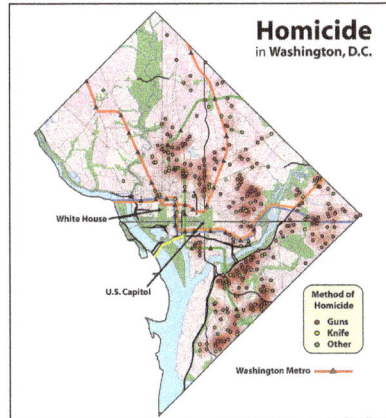

Crime Analysis: GIS is a necessary tool for crime mapping in law enforcement agencies worldwide. Crime mapping is a key component of crime analysis. Satellite images can display important information about criminal activities. The efficiency and the speed of the GIS analysis will increase the capabilities of crime fighting.

River Crossing Site Selection for Bridges: The important geotechnical consideration is the stability of slope leading down to and up from the water crossing. It is advisable to collect historical data on erosion and sedimentation. On the basis of these information asses the amount of river channel contraction, degree of curvature of river bend, nature of bed and bank materials including the flood flow and the flow depth, all these can be done in GIS within estimated time and accurately. This information has been often used for river crossing site selection for bridges.

Land Use Changes Associated with Open Cast Strip Mining: Mining is the back bone of the developing economy of any country. Mapping, monitoring and controlling the impact caused by the mining activities is necessary so as to understand the character and magnitude of these hazardous events in an area. The data required to understand the impact of mining from the environment is coming from different discipline, which need integration in order to arrive hazard map zonation.

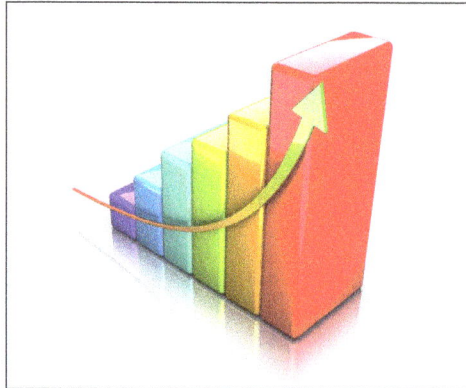

Economic Development: GIS technology is a valuable tool used for the economic development. It helps in site selection, suitability analysis, and for finding the right sites to locate new business and grow existing ones. Within economic development, GIS is used to support the emerging trend of economic gardening, a new way to foster local and regional economic growth by existing small business in the community.

School Student Walking Distance Analysis: If your community buses students to school, but only if they lived beyond a certain distance from their school, this can be used to determine what addresses are eligible for busing.

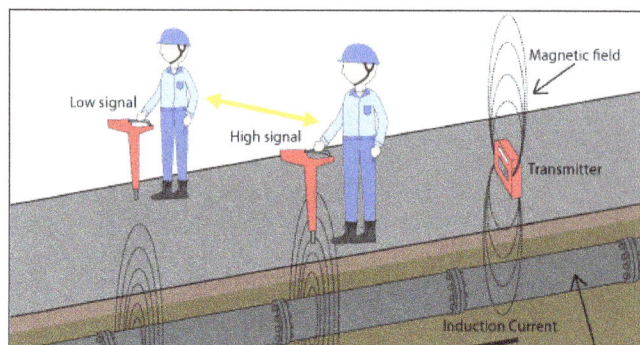

Locating Underground Pipes and Cables: Pipe line and cable location is essential for leak detection. It can be used to understand your water network, conducting repairs and adjustments, locating leaks known distance for correlating etc. Pipelines are continually monitored, check for leak detection and avoid the problem of geo hazards.

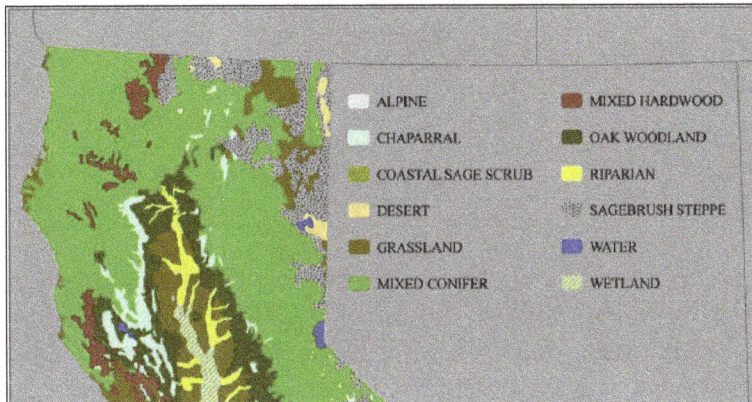

Coastal Vegetation Mapping and Conservation: Coastal vegetation like Mangroves are the protectors of coast from natural hazards like tsunami , so that the conservation of these vegetation are highly important. GIS enable us to map which are having higher density of vegetation and which area need more vegetated. Integration of these details to coastal zone mapping helps to identify the area prone to coastal erosion and we can plant more vegetation to reduce coastal erosion.

Regional Planning: Every day, planners use Geographic Information System (GIS) technology to research, develop, implement, and monitor the progress of their plans. GIS provides planners, surveyors, and engineers with the tools they need to design and map their neighborhoods and cities. Planners have the technical expertise, political savvy, and fiscal understanding to transform a vision of tomorrow into a strategic action plan for today, and they use GIS to facilitate the decision-making process.

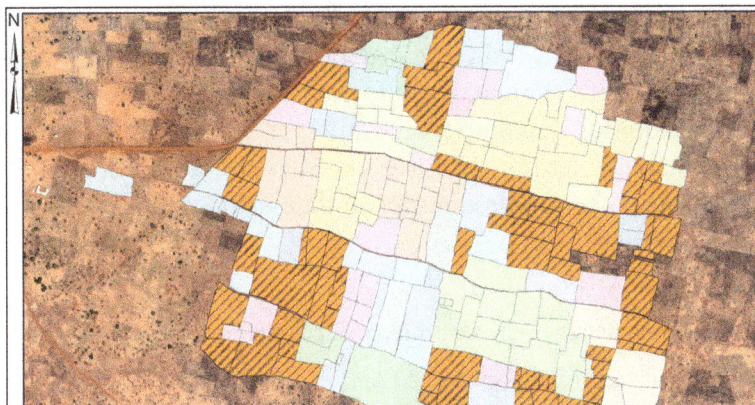

GIS for Land Administration: In a number of countries, the separate functions of land administration are being drawn together through the creation of digital cadastral databases, with these database they can reuse land for suitable needs, digital taxation and even utilities are also easily handle using these database.

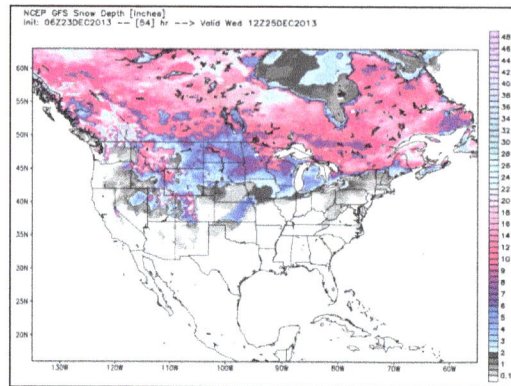

Snow Cover Mapping and Runoff Prediction: Systematic, periodical and precise snow cover mapping supported by GIS technology, and the organization of the results in a snow cover information system forms the basis for a wide range of applications. On the practical side, these applications are related to the monitoring of seasonal and yearly alterations of the snow cover under the presently existing climatic conditions, to simulate and forecast runoff, to map the regional distribution of the water equivalent, and to document the recession process of the snow cover during the melting period in its relation to geological features.

GIS for Wildlife Management: Man made destruction such as habitat loss, pollution, invasive species introduction, and climate change, are all threats to wildlife health and biodiversity. GIS technology is an effective tool for managing, analyzing, and visualizing wildlife data to target areas where international management practices are needed and to monitor their effectiveness. GIS helps wildlife management professionals examine and envision.

References

* Geographic-information-system-gis, encyclopedia: nationalgeographic.org, Retrieved 15 June, 2019

* Gis-applications-uses: grindgis.com, Retrieved 04 April, 2019

* Hager, Georg; Wellein, Gerhard (2010). Introduction to High Performance Computing for Scientists and Engineers. Chapman & Hall/CRC Computational Science. 20102232.doi:10.1201/EBK1439811924. ISBN 978-1-4398-1192-4. ISSN 2154-4492.

* Anselin, Luc; Rey, Serge (2012). "Spatial Econometrics in an Age of CyberGIScience". International Journal of Geographical Information Science. 26 (12): 2211–2226. doi:10.1080/13658816.2012.664276

6
Allied Fields

Topography is an interdisciplinary subject plays vital roles in many other fields. Some of them are earth sciences, planetary science, geography, geology, geomorphology, geoscience, etc. This chapter has been carefully written to provide an in-depth understanding of the varied facets of these allied fields of topography.

Earth Sciences

Earth sciences are the fields of study concerned with the solid Earth, its waters, and the air that envelops it. Included are the geologic, hydrologic, and atmospheric sciences.

The broad aim of the Earth sciences is to understand the present features and the past evolution of Earth and to use this knowledge, where appropriate, for the benefit of humankind. Thus, the basic concerns of the Earth scientist are to observe, describe, and classify all the features of the Earth, whether characteristic or not, to generate hypotheses with which to explain their presence and their development, and to devise means of checking opposing ideas for their relative validity. In this way the most plausible, acceptable, and long-lasting ideas are developed.

The physical environment in which humans live includes not only the immediate surface of the solid Earth but also the ground beneath it and the water and air above it. Early humans were more involved with the practicalities of life than with theories, and, thus, their survival depended on their ability to obtain metals from the ground to produce, for example, alloys, such as bronze from copper and tin, for tools and armour, to find adequate water supplies for establishing dwelling sites, and to forecast the weather, which had a far greater bearing on human life in earlier times than it has today. Such situations represent the foundations of the three principal component disciplines of the modern Earth sciences.

The rapid development of science as a whole over the past century and a half has given rise to an immense number of specializations and subdisciplines, with the result that the modern Earth scientist, perhaps unfortunately, tends to know a great deal about a very small area of study but only a little about most other aspects of the entire field. It is therefore very important for the layperson and the researcher alike to be aware of the complex interlinking network of disciplines that make up the Earth sciences today. Only when one is aware of the marvelous complexity of the Earth sciences and yet can understand the breakdown of the component disciplines is one in a position to select those parts of the subject that are of greatest personal interest.

It is worth emphasizing two important features that the three divisions of the Earth sciences have in common. First is the inaccessibility of many of the objects of study. Many rocks, as well as water and oil reservoirs, are at great depths in the Earth, while air masses circulate at vast heights above it. Thus, the Earth scientist has to have a good three-dimensional perspective. Second, there is the fourth dimension: time. The Earth scientist is responsible for working out how the Earth evolved over millions of years. For example, what were the physical and chemical conditions operating on the Earth and the Moon 3.5 billion years ago? How did the oceans form, and how did their chemical composition change with time? How has the atmosphere developed? And finally, How did life on Earth begin? and From what did humankind evolve?

Today the Earth sciences are divided into many disciplines, which are themselves divisible into six groups:

1. Those subjects that deal with the water and air at or above the solid surface of the Earth. These include the study of the water on and within the ground (hydrology), the glaciers and ice caps (glaciology), the oceans (oceanography), the atmosphere and its phenomena (meteorology), and the world's climates (climatology). In this part such fields of study are grouped under the hydrologic and atmospheric sciences and are treated separately from the geologic sciences, which focus on the solid Earth.

2. Disciplines concerned with the physical-chemical makeup of the solid Earth, which include the study of minerals (mineralogy), the three main groups of rocks (igneous, sedimentary, and metamorphic petrology), the chemistry of rocks (geochemistry), the structures in rocks (structural geology), and the physical properties of rocks at the Earth's surface and in its interior (geophysics).

3. The study of landforms (geomorphology), which is concerned with the description of the features of the present terrestrial surface and an analysis of the processes that gave rise to them.

4. Disciplines concerned with the geologic history of the Earth, including the study of fossils and the fossil record (paleontology), the development of sedimentary strata deposited typically over millions of years (stratigraphy), and the isotopic chemistry and age dating of rocks (geochronology).

5. Applied Earth sciences dealing with current practical applications beneficial to society. These include the study of fossil fuels (oil, natural gas, and coal); oil reservoirs; mineral deposits; geothermal energy for electricity and heating; the structure and composition of bedrock for the location of bridges, nuclear reactors, roads, dams, and skyscrapers and other buildings; hazards involving rock and mud avalanches, volcanic eruptions, earthquakes, and the collapse of tunnels; and coastal, cliff, and soil erosion.

6. The study of the rock record on the Moon and the planets and their satellites (astrogeology). This field includes the investigation of relevant terrestrial features—namely, tektites (glassy objects resulting from meteorite impacts) and astroblemes (meteorite craters).

With such intergradational boundaries between the divisions of the Earth sciences (which, on a broader scale, also intergrade with physics, chemistry, biology, mathematics, and certain branches of engineering), researchers today must be versatile in their approach to problems. Hence, an important aspect of training within the Earth sciences is an appreciation of their multidisciplinary nature.

Planetary Science

Planetary science or, more rarely, planetology, is the scientific study of planets (including Earth), moons, and planetary systems (in particular those of the Solar System) and the processes that form them. It studies objects ranging in size from micrometeoroids to gas giants, aiming to determine their composition, dynamics, formation, interrelations and history. It is a strongly interdisciplinary field, originally growing from astronomy and earth science, but which now incorporates many disciplines, including planetary geology (together with geochemistry and geophysics), cosmochemistry, atmospheric science, oceanography, hydrology, theoretical planetary science, glaciology, and exoplanetology. Allied disciplines include space physics, when concerned with the effects of the Sun on the bodies of the Solar System, and astrobiology.

There are interrelated observational and theoretical branches of planetary science. Observational research can involve a combination of space exploration, predominantly with robotic spacecraft missions using remote sensing, and comparative, experimental work in Earth-based laboratories. The theoretical component involves considerable computer simulation and mathematical modelling.

Planetary scientists are generally located in the astronomy and physics or Earth sciences departments of universities or research centres, though there are several purely planetary science institutes worldwide. Some planetary scientists work at private research centres and often initiate partnership research tasks.

Disciplines

Planetary Astronomy

This is both an observational and a theoretical science. Observational researchers are predominantly concerned with the study of the small bodies of the Solar System: those that are observed by telescopes, both optical and radio, so that characteristics of these bodies such as shape, spin, surface materials and weathering are determined, and the history of their formation and evolution can be understood.

Theoretical planetary astronomy is concerned with dynamics: the application of the principles of celestial mechanics to the Solar System and extrasolar planetary systems.

Planetary Geology

The best known research topics of planetary geology deal with the planetary bodies in the near vicinity of the Earth: the Moon, and the two neighbouring planets: Venus and Mars. Of these, the Moon was studied first, using methods developed earlier on the Earth.

Geomorphology

Geomorphology studies the features on planetary surfaces and reconstructs the history of their formation, inferring the physical processes that acted on the surface. Planetary geomorphology includes the study of several classes of surface features:

- Impact features (multi-ringed basins, craters).

- Volcanic and tectonic features (lava flows, fissures, rilles).

- Space weathering - erosional effects generated by the harsh environment of space (continuous micro meteorite bombardment, high-energy particle rain, impact gardening). For example, the thin dust cover on the surface of the lunar regolith is a result of micro meteorite bombardment.

- Hydrological features: the liquid involved can range from water to hydrocarbon and ammonia, depending on the location within the Solar System.

The history of a planetary surface can be deciphered by mapping features from top to bottom according to their deposition sequence, as first determined on terrestrial strata by Nicolas Steno. For example, stratigraphic mapping prepared the Apollo astronauts for the field geology they would encounter on their lunar missions. Overlapping sequences were identified on images taken by the Lunar Orbiter program, and these were used to prepare a lunar stratigraphic column and geological map of the Moon.

Cosmochemistry, Geochemistry and Petrology

One of the main problems when generating hypotheses on the formation and evolution of objects in the Solar System is the lack of samples that can be analysed in the laboratory, where a large suite of tools are available and the full body of knowledge derived from terrestrial geology can be brought to bear. Direct samples from the Moon, asteroids and Mars are present on Earth, removed from their parent bodies and delivered as meteorites. Some of these have suffered contamination from the oxidising effect of Earth's atmosphere and the infiltration of the biosphere, but those meteorites collected in the last few decades from Antarctica are almost entirely pristine.

The different types of meteorites that originate from the asteroid belt cover almost all parts of the structure of differentiated bodies: meteorites even exist that come from the core-mantle boundary (pallasites). The combination of geochemistry and observational astronomy has also made it possible to trace the HED meteorites back to a specific asteroid in the main belt, 4 Vesta.

The comparatively few known Martian meteorites have provided insight into the geochemical composition of the Martian crust, although the unavoidable lack of information about their points of origin on the diverse Martian surface has meant that they do not provide more detailed constraints on theories of the evolution of the Martian lithosphere. As of July 24, 2013 65 samples of Martian meteorites have been discovered on Earth. Many were found in either Antarctica or the Sahara Desert.

During the Apollo era, in the Apollo program, 384 kilograms of lunar samples were collected and transported to the Earth, and 3 Soviet Luna robots also delivered regolith samples from the Moon.

These samples provide the most comprehensive record of the composition of any Solar System body beside the Earth. The numbers of lunar meteorites are growing quickly in the last few years – as of April 2008 there are 54 meteorites that have been officially classified as lunar. Eleven of these are from the US Antarctic meteorite collection, 6 are from the Japanese Antarctic meteorite collection, and the other 37 are from hot desert localities in Africa, Australia, and the Middle East. The total mass of recognized lunar meteorites is close to 50 kg.

Geophysics

Space probes made it possible to collect data in not only the visible light region, but in other areas of the electromagnetic spectrum. The planets can be characterized by their force fields: gravity and their magnetic fields, which are studied through geophysics and space physics.

Measuring the changes in acceleration experienced by spacecraft as they orbit has allowed fine details of the gravity fields of the planets to be mapped. For example, in the 1970s, the gravity field disturbances above lunar maria were measured through lunar orbiters, which led to the discovery of concentrations of mass, mascons, beneath the Imbrium, Serenitatis, Crisium, Nectaris and Humorum basins.

The solar wind is deflected by the magnetosphere (not to scale).

If a planet's magnetic field is sufficiently strong, its interaction with the solar wind forms a magnetosphere around a planet. Early space probes discovered the gross dimensions of the terrestrial magnetic field, which extends about 10 Earth radii towards the Sun. The solar wind, a stream of charged particles, streams out and around the terrestrial magnetic field, and continues behind the magnetic tail, hundreds of Earth radii downstream. Inside the magnetosphere, there are relatively dense regions of solar wind particles, the Van Allen radiation belts.

Geophysics includes seismology and tectonophysics, geophysical fluid dynamics, mineral physics, geodynamics, mathematical geophysics, and geophysical surveying.

Planetary geodesy, (also known as planetary geodetics) deals with the measurement and representation of the planets of the Solar System, their gravitational fields and geodynamic phenomena (polar motion in three-dimensional, time-varying space. The science of geodesy has elements of both astrophysics and planetary sciences. The shape of the Earth is to a large extent the result of its rotation, which causes its equatorial bulge, and the competition of geologic processes such as the collision of plates and of vulcanism, resisted by the Earth's gravity field. These principles can be applied to the solid surface of Earth (orogeny). Few mountains are higher than 10 km (6 mi), few deep sea trenches deeper than that because quite simply, a mountain as tall as, for example, 15 km (9 mi), would develop so much pressure at its base, due to gravity, that the rock there would become plastic, and the mountain would slump back to a height of roughly 10 km (6 mi) in a geologically insignificant time. Some or all of these geologic principles can be applied to other planets besides Earth. For instance on Mars, whose surface gravity is much less, the largest volcano, Olympus Mons, is 27 km (17 mi) high at its peak, a height that could not be maintained on Earth. The Earth geoid is essentially the figure of the Earth abstracted from its topographic features. Therefore, the Mars geoid is essentially the figure of Mars abstracted from its topographic features. Surveying and mapping are two important fields of application of geodesy.

Atmospheric Science

Cloud bands clearly visible on Jupiter.

The atmosphere is an important transitional zone between the solid planetary surface and the higher rarefied ionizing and radiation belts. Not all planets have atmospheres: their existence depends on the mass of the planet, and the planet's distance from the Sun — too distant and frozen atmospheres occur. Besides the four gas giant planets, almost all of the terrestrial planets (Earth, Venus, and Mars) have significant atmospheres. Two moons have significant atmospheres: Saturn's moon Titan and Neptune's moon Triton. A tenuous atmosphere exists around Mercury.

The effects of the rotation rate of a planet about its axis can be seen in atmospheric streams and currents. Seen from space, these features show as bands and eddies in the cloud system, and are particularly visible on Jupiter and Saturn.

Comparative Planetary Science

Planetary science frequently makes use of the method of comparison to give a greater understanding of the object of study. This can involve comparing the dense atmospheres of Earth and Saturn's moon Titan, the evolution of outer Solar System objects at different distances from the Sun, or the geomorphology of the surfaces of the terrestrial planets, to give only a few examples.

The main comparison that can be made is to features on the Earth, as it is much more accessible and allows a much greater range of measurements to be made. Earth analogue studies are particularly common in planetary geology, geomorphology, and also in atmospheric science.

Geography

Geography is the study of the Earth's physical features and environment including the impact of human activity on these factors and vice versa. The subject also encompasses the study of patterns of human population distribution, land use, resource availability, and industries.

Scholars who study geography are known as geographers. These people engage themselves in the exciting task of exploring and studying the Earth's natural environment and human society. Although map-makers were known as geographers in the ancient world, today, they are more specifically known as cartographers. Geographers usually focus on two major fields of geographical studies: physical geography or human geography.

The term geography was coined by the ancient Greeks who not only created detailed maps and accounts of places around them but also illuminated why and how human and natural patterns varied from one place to another on Earth. Through the passage of time, the rich legacy of geography made a momentous journey to the bright Islamic minds. The Islamic Golden Age witnessed astounding advancements in the geographical sciences. Islamic geographers were credited with groundbreaking discoveries. New lands were explored and the world's first grid-based mapping system was developed. The Chinese civilization also contributed instrumentally towards the development of early geography. The compass, a traveling aid, devised by the Chinese, was used by the Chinese explorers to explore the unknown.

A new historical chapter of geography opened during the "Age of Discovery", a period coinciding with the European Renaissance. A fresh interest in geography was regenerated in the European world. Marco Polo, the Venetian merchant traveler, spearheaded this new Age of Exploration. Commercial interests in establishing trade contacts with the rich civilizations of Asia like China and India became the primary reason for traveling during this period. Europeans moved ahead in all directions, discovering new lands, unique cultures, and natural wonders in the process. They also began to colonize new lands towards the latter half of the Age of Exploration. The tremendous potential of geography to shape the future of human civilization was recognized and in the 18th Century, geography was introduced as a discipline of study at the university level. Based on geographical knowledge, the human society discovered new ways and means to overcome the challenges posed by nature and human civilizations flourished in all parts of the world. In the 20th century, aerial photography, satellite technology, computerized systems, and sophisticated software radically changed the definition of geography and made the study of geography more comprehensive and detailed.

The Branches of Geography

Geography can be regarded as an interdisciplinary science. The subject encompasses an interdisciplinary perspective that allows the observation and analysis of anything distributed in Earth space and the development of solutions to problems based on such analysis. The discipline of geography can be divided into several branches of study. The primary classification of geography divides the approach to the subject into the two broad categories of physical geography and human geography.

Physical Geography

Physical geography is defined as the branch of geography that encompasses the study of the natural features and phenomena (or processes) on the Earth.

Physical geography may be further subdivided into various branches:

1. Geomorphology: This involves the study of the topographic and bathymetric features on Earth. The science helps to elucidate various aspects related to the landforms on the Earth such as their history and dynamics. Geomorphology also attempts to predict the future changes in the Earth's physical features.

2. Glaciology: This field of physical geography deals with the study of the inter-dynamics of glaciers and their effects on the planet's environment. Thus, glaciology involves the study of the cryosphere including the alpine glaciers and the continental glaciers. Glacial geology, snow hydrology, etc., are some of the sub-fields of glaciological studies.

3. Oceanography: Since oceans hold 96.5% of the Earth's waters, a special field of oceanography needs to be dedicated to the study of oceans. The science of oceanography includes geological oceanography (study of the geological aspects of the ocean floor, its mountains, volcanoes, etc.), biological oceanography (study of the marine life and ocean ecosystems), chemical oceanography (study of the chemical composition of the marine waters and their effects on marine life forms), physical oceanography (study of the oceanic movements like the waves, currents, etc.).

4. Hydrology: This is another vital aspect of physical geography. Hydrology deals with the study of the properties of the Earth's water resources and the movement dynamics of water in relation to land. The field encompasses the study of the rivers, lakes, glaciers, and underground aquifers on the planet. It studies the continuous movement of water from one source to another on, above, and below the Earth's surface, in the form of the hydrological cycle.

5. Pedology: A branch of soil science, pedology involves the study of the different soil types in their natural environment on the surface of the Earth. This field of study helps gather information and knowledge on the process of soil formation (pedogenesis), soil constitution, soil texture, classification, etc.

6. Biogeography: An indispensable field of physical geography, biogeography is the study of how species on Earth are dispersed in geographic space. It also deals with the distribution of species over geological time periods. Each geographical area has its own unique ecosystem and biogeography explores and explains such ecosystems in relation to physical geographical features. Different branches of biogeography exist like zoogeography (geographic distribution of animals), phytogeography (geographic distribution of plants), insular biogeography (the study of factors influencing isolated ecosystems), etc.

7. Paleogeography: This branch of physical geography examines the geographical features at various time points in the Earth's geological history. It helps the geographers to attain knowledge about the continental positions and plate tectonics determined by studying paleomagnetism and fossil records.

8. Climatology: The scientific study of climate, climatology is a crucial field of geographical studies in today's world. It examines all aspects related to the micro or local climates of places and also the macro or global climate. It also involves an examination of the impact of human society on climate and vice versa.

9. Meteorology: This field of physical geography is concerned with the study of the weather patterns of a place and the atmospheric processes and phenomena that influence the weather.

10. Environmental geography: Also known as integrative geography, this field of physical geography explores the interactions between humans (individuals or society) and their natural environment from the spatial point of view. Environmental geography is thus the bridging gap between human geography and physical geography and can be treated as an amalgamation of multiple fields of physical geography and human geography.

11. Coastal geography: Coastal geography is another area of specialization of physical geography that also involves a study of human geography. It deals with the study of the dynamic interface between the coastal land and the sea. The physical processes that shape the coastal landscape and the influence of the sea in triggering landscape modifications is incorporated in the study of coastal geography. The study also involves an understanding of the ways the human inhabitants of coastal areas influence the coastal landforms and ecosystems.

12. Quaternary science: This is a highly specialized field of physical geography that deals with the study of the Quaternary period on Earth (the Earth's geographical history encompassing the last 2.6 million years). It allows the geographers to learn about the environmental changes undergone in the planet's recent past. This knowledge is then used as a tool to predict future changes in the Earth's environment.

13. Geomatics: Geomatics is a technical branch of physical geography that involves the collection of data related to the earth's surface, analysis of the data, its interpretation, and storage. Geodesy, remote sensing, and geographical information science are the three sub-divisions of geomatics.

14. Landscape ecology: The science of landscape ecology deals with the study of how the varying landscapes on Earth influences the ecological processes and ecosystems on the planet. The German geographer Carl Troll is credited as the founder of this field of physical geography.

Human Geography

Human geography is the branch of geography that deals with the study of how the human society is influenced by the Earth's surface and environment and how, in turn, anthropological activities impact the planet. Human geography is centered on the study of the planet's most evolved creatures: the humans and their environment.

Geology

Geology is an earth science concerned with the solid Earth, the rocks of which it is composed, and the processes by which they change over time. Geology can also include the study of the solid features of any terrestrial planet or natural satellite such as Mars or the Moon. Modern geology significantly overlaps all other earth sciences, including hydrology and the atmospheric sciences, and so is treated as one major aspect of integrated earth system science and planetary science.

Geology describes the structure of the Earth on and beneath its surface, and the processes that have shaped that structure. It also provides tools to determine the relative and absolute ages of rocks found in a given location, and also to describe the histories of those rocks. By combining these tools, geologists are able to chronicle the geological history of the Earth as a whole, and also to demonstrate the age of the Earth. Geology provides the primary evidence for plate tectonics, the evolutionary history of life, and the Earth's past climates.

An 1875 geological map of Europe, compiled by the Belgian geologist André Dumont (colors indicate the distribution of rocks of different ages and types across the continent, as they were known then).

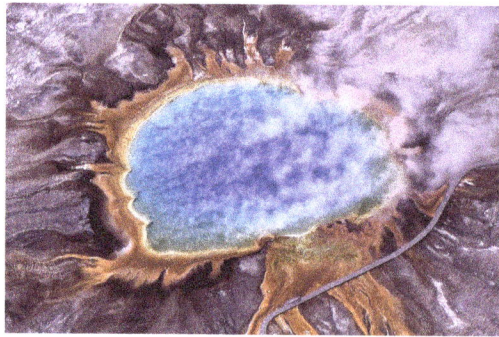

Aerial view of Grand Prismatic Spring; Hot Springs, Midway & Lower Geyser Basin, Yellowstone National Park.

Geologists use a wide variety of methods to understand the Earth's structure and evolution, including field work, rock description, geophysical techniques, chemical analysis, physical experiments, and numerical modelling. In practical terms, geology is important for mineral and hydrocarbon exploration and exploitation, evaluating water resources, understanding of natural hazards, the remediation of environmental problems, and providing insights into past climate change. Geology is a major academic discipline, and it plays an important role in geotechnical engineering.

Kinney Lake and Mount Whitehorn near Mount Robson, Canada.

Sarychev Peak Volcano erupts on Matua Island.

Geologic Materials

The majority of geological data comes from research on solid Earth materials. These typically fall into one of two categories: rock and unlithified material.

Rock

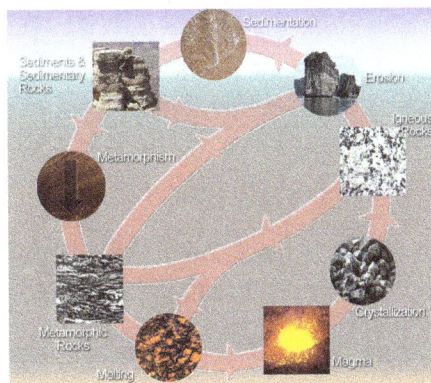

The rock cycle shows the relationship between igneous,
sedimentary, and metamorphic rocks.

The majority of research in geology is associated with the study of rock, as rock provides the primary record of the majority of the geologic history of the Earth. There are three major types of rock: igneous, sedimentary, and metamorphic. The rock cycle illustrates the relationships among them.

When a rock solidifies or crystallizes from melt (magma or lava), it is an igneous rock. This rock can be weathered and eroded, then redeposited and lithified into a sedimentary rock. It can then be turned into a metamorphic rock by heat and pressure that change its mineral content, resulting in a characteristic fabric. All three types may melt again, and when this happens, new magma is formed, from which an igneous rock may once more solidify.

Native gold from Venezuela.

Quartz from Tibet.

Tests

To study all three types of rock, geologists evaluate the minerals of which they are composed. Each mineral has distinct physical properties, and there are many tests to determine each of them. The specimens can be tested for:

- Luster: Measurement of the amount of light reflected from the surface. Luster is broken into metallic and nonmetallic.

- Color: Minerals are grouped by their color. Mostly diagnostic but impurities can change a mineral's color.

- Streak: Performed by scratching the sample on a porcelain plate. The color of the streak can help name the mineral.

- Hardness: The resistance of a mineral to scratch.

- Breakage pattern: A mineral can either show fracture or cleavage, the former being breakage of uneven surfaces and the latter a breakage along closely spaced parallel planes.

- Specific gravity: The weight of a specific volume of a mineral.

- Effervescence: Involves dripping hydrochloric acid on the mineral to test for fizzing.

- Magnetism: Involves using a magnet to test for magnetism.

- Taste: Minerals can have a distinctive taste, like Halite (mineral) (which tastes like table salt).

- Smell: Minerals can have a distinctive odor. For example, sulfur smells like rotten eggs.

Unlithified Material

Geologists also study unlithified materials, which typically come from more recent deposits. These materials are superficial deposits that lie above the bedrock. This study is often known as Quaternary geology, after the Quaternary period of geologic history.

Magma and Lava

However, unlithified material does not only include sediments. Magmas and lavas are the original unlithified source of all igneous rocks. The active flow of molten rock is closely studied in volcanology, and igneous petrology aims to determine the history of igneous rocks from their final crystallization to their original molten source.

Whole-earth Structure

Plate Tectonics

In the 1960s, it was discovered that the Earth's lithosphere, which includes the crust and rigid uppermost portion of the upper mantle, is separated into tectonic plates that move across the plastically deforming, solid, upper mantle, which is called the asthenosphere. This theory is supported by several types of observations, including seafloor spreading and the global distribution of mountain terrain and seismicity.

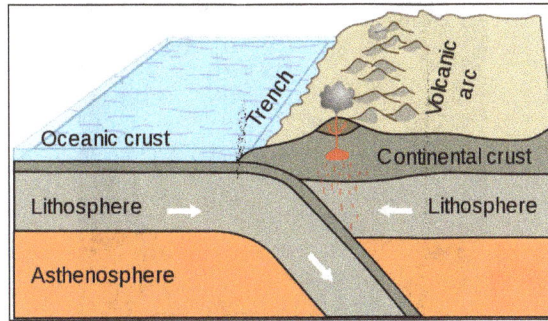

Oceanic-continental convergence resulting in subduction and
volcanic arcs illustrates one effect of plate tectonics.

There is an intimate coupling between the movement of the plates on the surface and the convection of the mantle (that is, the heat transfer caused by bulk movement of molecules within fluids). Thus, oceanic plates and the adjoining mantle convection currents always move in the same direction – because the oceanic lithosphere is actually the rigid upper thermal boundary layer of the convecting mantle. This coupling between rigid plates moving on the surface of the Earth and the convecting mantle is called plate tectonics.

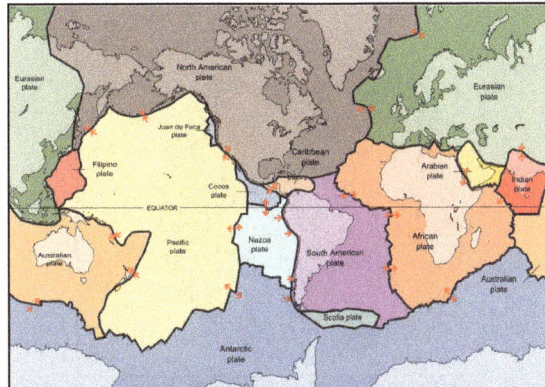

The major tectonic plates of the Earth.

The development of plate tectonics has provided a physical basis for many observations of the solid Earth. Long linear regions of geologic features are explained as plate boundaries.

For example:

- Mid-ocean ridges, high regions on the seafloor where hydrothermal vents and volcanoes exist, are seen as divergent boundaries, where two plates move apart.

- Arcs of volcanoes and earthquakes are theorized as convergent boundaries, where one plate subducts, or moves, under another.

Transform boundaries, such as the San Andreas Fault system, resulted in widespread powerful earthquakes. Plate tectonics also has provided a mechanism for Alfred Wegener's theory of continental drift, in which the continents move across the surface of the Earth over geologic time. They also provided a driving force for crustal deformation, and a new setting for the observations of structural geology. The power of the theory of plate tectonics lies in its ability to combine all of these observations into a single theory of how the lithosphere moves over the convecting mantle.

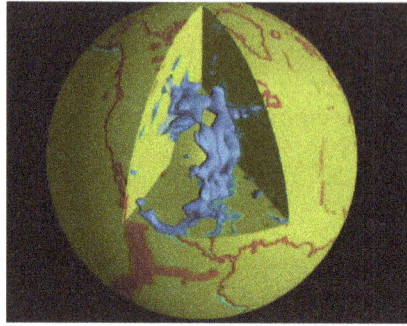

In this diagram based on seismic tomography, subducting slabs
are in blue and continental margins and a few plate boundaries are in red.

The blue blob in the cutaway section is the Farallon Plate, which is subducting beneath North
America. The remnants of this plate on the surface of the Earth are the Juan de Fuca Plate and
Explorer Plate, both in the northwestern United States and southwestern Canada, and the Cocos
Plate on the west coast of Mexico.

Earth Structure

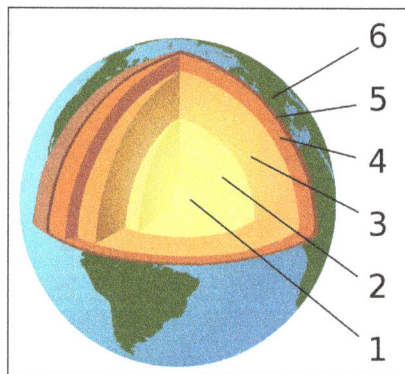

The Earth's layered structure. (1) inner core; (2) outer core; (3) lower mantle;
(4) upper mantle; (5) lithosphere; (6) crust (part of the lithosphere).

Advances in seismology, computer modeling, and mineralogy and crystallography at high temperatures and pressures give insights into the internal composition and structure of the Earth.

Seismologists can use the arrival times of seismic waves in reverse to image the interior of the
Earth. Early advances in this field showed the existence of a liquid outer core (where shear waves
were not able to propagate) and a dense solid inner core. These advances led to the development
of a layered model of the Earth, with a crust and lithosphere on top, the mantle below (separated
within itself by seismic discontinuities at 410 and 660 kilometers), and the outer core and inner
core below that. More recently, seismologists have been able to create detailed images of wave
speeds inside the earth in the same way a doctor images a body in a CT scan. These images have led
to a much more detailed view of the interior of the Earth, and have replaced the simplified layered
model with a much more dynamic model.

Mineralogists have been able to use the pressure and temperature data from the seismic and
modelling studies alongside knowledge of the elemental composition of the Earth to reproduce
these conditions in experimental settings and measure changes in crystal structure. These studies

explain the chemical changes associated with the major seismic discontinuities in the mantle and show the crystallographic structures expected in the inner core of the Earth.

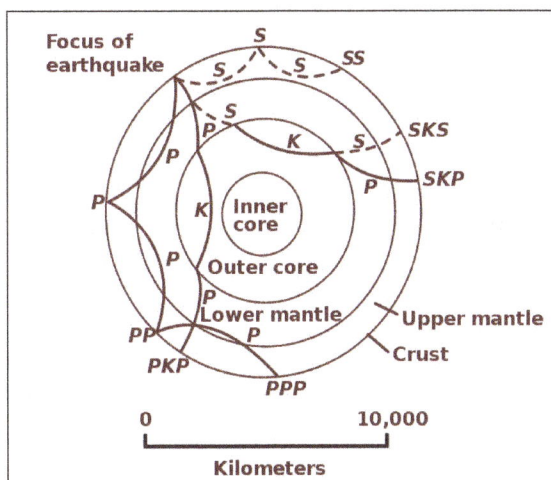

Earth layered structure. Typical wave paths from earthquakes like these gave early seismologists insights into the layered structure of the Earth.

Dating Methods

Relative Dating

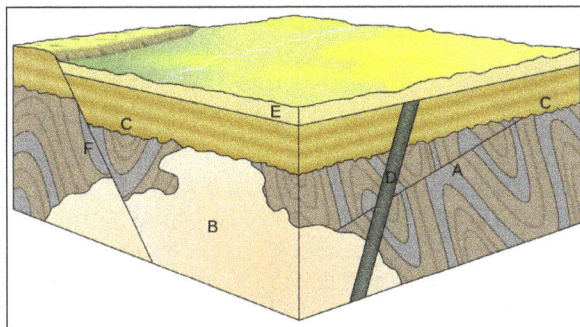

Cross-cutting relations can be used to determine the relative ages of rock strata and other geological structures.

Explanations: A – folded rock strata cut by a thrust fault; B – large intrusion (cutting through A); C – erosional angular unconformity (cutting off A & B) on which rock strata were deposited; D – volcanic dyke (cutting through A, B & C); E – even younger rock strata (overlying C & D); F – normal fault (cutting through A, B, C & E).

Methods for relative dating were developed when geology first emerged as a natural science. Geologists still use the following principles today as a means to provide information about geologic history and the timing of geologic events.

The principle of uniformitarianism states that the geologic processes observed in operation that modify the Earth's crust at present have worked in much the same way over geologic time. A fundamental principle of geology advanced by the 18th century Scottish physician and geologist James Hutton is that "the present is the key to the past." In Hutton's words: "the past history of our globe must be explained by what can be seen to be happening now."

The principle of intrusive relationships concerns crosscutting intrusions. In geology, when an igneous intrusion cuts across a formation of sedimentary rock, it can be determined that the igneous intrusion is younger than the sedimentary rock. Different types of intrusions include stocks, laccoliths, batholiths, sills and dikes.

The principle of cross-cutting relationships pertains to the formation of faults and the age of the sequences through which they cut. Faults are younger than the rocks they cut; accordingly, if a fault is found that penetrates some formations but not those on top of it, then the formations that were cut are older than the fault, and the ones that are not cut must be younger than the fault. Finding the key bed in these situations may help determine whether the fault is a normal fault or a thrust fault.

The principle of inclusions and components states that, with sedimentary rocks, if inclusions (or *clasts*) are found in a formation, then the inclusions must be older than the formation that contains them. For example, in sedimentary rocks, it is common for gravel from an older formation to be ripped up and included in a newer layer. A similar situation with igneous rocks occurs when xenoliths are found. These foreign bodies are picked up as magma or lava flows, and are incorporated, later to cool in the matrix. As a result, xenoliths are older than the rock that contains them.

The Permian through Jurassic stratigraphy of the Colorado
Plateau area of southeastern Utah is an example of both
original horizontality and the law of superposition.

These strata make up much of the famous prominent rock formations in widely spaced protected areas such as Capitol Reef National Park and Canyonlands National Park. From top to bottom: Rounded tan domes of the Navajo Sandstone, layered red Kayenta Formation, cliff-forming, vertically jointed, red Wingate Sandstone, slope-forming, purplish Chinle Formation, layered, lighter-red Moenkopi Formation, and white, layered Cutler Formation sandstone. Picture from Glen Canyon National Recreation Area, Utah.

The principle of original horizontality states that the deposition of sediments occurs as essentially horizontal beds. Observation of modern marine and non-marine sediments in a wide variety of environments supports this generalization (although cross-bedding is inclined, the overall orientation of cross-bedded units is horizontal).

The principle of superposition states that a sedimentary rock layer in a tectonically undisturbed sequence is younger than the one beneath it and older than the one above it. Logically a younger layer cannot slip beneath a layer previously deposited. This principle allows sedimentary layers to be viewed as a form of vertical time line, a partial or complete record of the time elapsed from deposition of the lowest layer to deposition of the highest bed.

The principle of faunal succession is based on the appearance of fossils in sedimentary rocks. As organisms exist during the same period throughout the world, their presence or (sometimes) absence provides a relative age of the formations where they appear. Based on principles that William Smith laid out almost a hundred years before the publication of Charles Darwin's theory of evolution, the principles of succession developed independently of evolutionary thought. The principle becomes quite complex, however, given the uncertainties of fossilization, localization of fossil types due to lateral changes in habitat (facies change in sedimentary strata), and that not all fossils formed globally at the same time.

Absolute Dating

The mineral zircon is often used in radiometric dating.

Geologists also use methods to determine the absolute age of rock samples and geological events. These dates are useful on their own and may also be used in conjunction with relative dating methods or to calibrate relative methods.

At the beginning of the 20th century, advancement in geological science was facilitated by the ability to obtain accurate absolute dates to geologic events using radioactive isotopes and other methods. This changed the understanding of geologic time. Previously, geologists could only use fossils and stratigraphic correlation to date sections of rock relative to one another. With isotopic dates, it became possible to assign absolute ages to rock units, and these absolute dates could be applied to fossil sequences in which there was datable material, converting the old relative ages into new absolute ages.

For many geologic applications, isotope ratios of radioactive elements are measured in minerals that give the amount of time that has passed since a rock passed through its particular closure temperature, the point at which different radiometric isotopes stop diffusing into and out of the crystal lattice. These are used in geochronologic and thermochronologic studies. Common methods include uranium-lead dating, potassium-argon dating, argon-argon dating and uranium-thorium dating. These methods are used for a variety of applications. Dating of lava and volcanic ash layers found within a stratigraphic sequence can provide absolute age data for sedimentary rock units that do not contain radioactive isotopes and calibrate relative dating techniques. These methods can also be used to determine ages of pluton emplacement. Thermochemical techniques can be used to determine temperature profiles within the crust, the uplift of mountain ranges, and paleotopography.

Fractionation of the lanthanide series elements is used to compute ages since rocks were removed from the mantle.

Other methods are used for more recent events. Optically stimulated luminescence and cosmogenic radionuclide dating are used to date surfaces and/or erosion rates. Dendrochronology can also be used for the dating of landscapes. Radiocarbon dating is used for geologically young materials containing organic carbon.

Geological Development of an Area

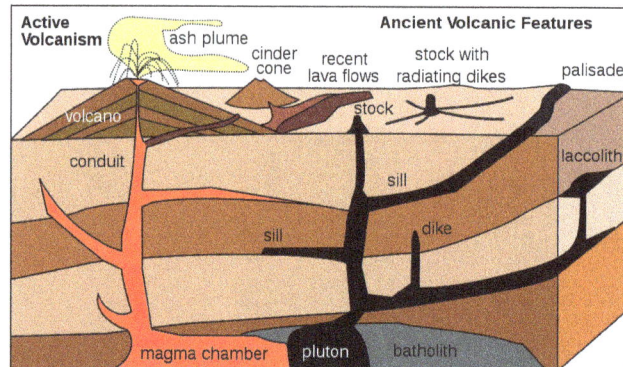

An originally horizontal sequence of sedimentary rocks
(in shades of tan) are affected by igneous activity.

Deep below the surface are a magma chamber and large associated igneous bodies. The magma chamber feeds the volcano, and sends offshoots of magma that will later crystallize into dikes and sills. Magma also advances upwards to form intrusive igneous bodies. The diagram illustrates both a cinder cone volcano, which releases ash, and a composite volcano, which releases both lava and ash.

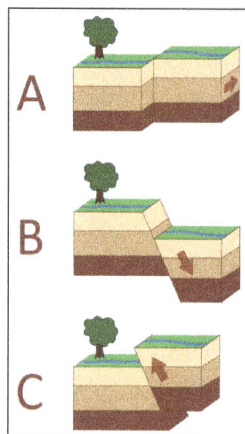

An illustration of the three types of faults:

a. Strike-slip faults occur when rock units slide past one another.

b. Normal faults occur when rocks are undergoing horizontal extension.

c. Reverse (or thrust) faults occur when rocks are undergoing horizontal shortening.

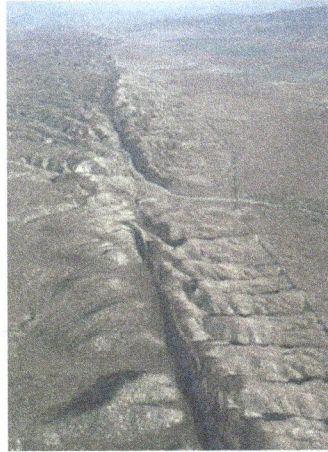

The San Andreas Fault in California.

The geology of an area changes through time as rock units are deposited and inserted, and deformational processes change their shapes and locations.

Rock units are first emplaced either by deposition onto the surface or intrusion into the overlying rock. Deposition can occur when sediments settle onto the surface of the Earth and later lithify into sedimentary rock, or when as volcanic material such as volcanic ash or lava flows blanket the surface. Igneous intrusions such as batholiths, laccoliths, dikes, and sills, push upwards into the overlying rock, and crystallize as they intrude.

After the initial sequence of rocks has been deposited, the rock units can be deformed and/or metamorphosed. Deformation typically occurs as a result of horizontal shortening, horizontal extension, or side-to-side (strike-slip) motion. These structural regimes broadly relate to convergent boundaries, divergent boundaries, and transform boundaries, respectively, between tectonic plates.

When rock units are placed under horizontal compression, they shorten and become thicker. Because rock units, other than muds, do not significantly change in volume, this is accomplished in two primary ways: through faulting and folding. In the shallow crust, where brittle deformation can occur, thrust faults form, which causes deeper rock to move on top of shallower rock. Because deeper rock is often older, as noted by the principle of superposition, this can result in older rocks moving on top of younger ones. Movement along faults can result in folding, either because the faults are not planar or because rock layers are dragged along, forming drag folds as slip occurs along the fault. Deeper in the Earth, rocks behave plastically and fold instead of faulting. These folds can either be those where the material in the center of the fold buckles upwards, creating "antiforms", or where it buckles downwards, creating "synforms". If the tops of the rock units within the folds remain pointing upwards, they are called anticlines and synclines, respectively. If some of the units in the fold are facing downward, the structure is called an overturned anticline or syncline, and if all of the rock units are overturned or the correct up-direction is unknown, they are simply called by the most general terms, antiforms and synforms.

Even higher pressures and temperatures during horizontal shortening can cause both folding and metamorphism of the rocks. This metamorphism causes changes in the mineral composition of the rocks; creates a foliation, or planar surface, that is related to mineral growth under stress. This can remove signs of the original textures of the rocks, such as bedding in sedimentary rocks, flow features of lavas, and crystal patterns in crystalline rocks.

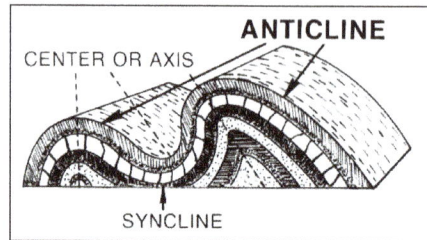

A diagram of folds, indicating an anticline and a syncline.

Extension causes the rock units as a whole to become longer and thinner. This is primarily accomplished through normal faulting and through the ductile stretching and thinning. Normal faults drop rock units that are higher below those that are lower. This typically results in younger units ending up below older units. Stretching of units can result in their thinning. In fact, at one location within the Maria Fold and Thrust Belt, the entire sedimentary sequence of the Grand Canyon appears over a length of less than a meter. Rocks at the depth to be ductilely stretched are often also metamorphosed. These stretched rocks can also pinch into lenses, known as boudins, after the French word for "sausage" because of their visual similarity.

Where rock units slide past one another, strike-slip faults develop in shallow regions, and become shear zones at deeper depths where the rocks deform ductilely.

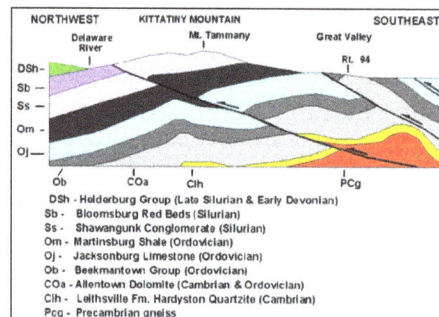

Geologic cross section of Kittatinny Mountain.

This cross section shows metamorphic rocks, overlain by younger sediments deposited after the metamorphic event. These rock units were later folded and faulted during the uplift of the mountain.

The addition of new rock units, both depositionally and intrusively, often occurs during deformation. Faulting and other deformational processes result in the creation of topographic gradients, causing material on the rock unit that is increasing in elevation to be eroded by hillslopes and channels. These sediments are deposited on the rock unit that is going down. Continual motion along the fault maintains the topographic gradient in spite of the movement of sediment, and continues to create accommodation space for the material to deposit. Deformational events are often also associated with volcanism and igneous activity. Volcanic ashes and lavas accumulate on the surface, and igneous intrusions enter from below. Dikes, long, planar igneous intrusions, enter along cracks, and therefore often form in large numbers in areas that are being actively deformed. This can result in the emplacement of dike swarms, such as those that are observable across the Canadian shield, or rings of dikes around the lava tube of a volcano.

All of these processes do not necessarily occur in a single environment, and do not necessarily occur in a single order. The Hawaiian Islands, for example, consist almost entirely of layered basaltic lava flows. The sedimentary sequences of the mid-continental United States and the Grand Canyon in the southwestern United States contain almost-undeformed stacks of sedimentary rocks that have remained in place since Cambrian time. Other areas are much more geologically complex. In the southwestern United States, sedimentary, volcanic, and intrusive rocks have been metamorphosed, faulted, foliated, and folded. Even older rocks, such as the Acasta gneiss of the Slave craton in northwestern Canada, the oldest known rock in the world have been metamorphosed to the point where their origin is undiscernable without laboratory analysis. In addition, these processes can occur in stages. In many places, the Grand Canyon in the southwestern United States being a very visible example, the lower rock units were metamorphosed and deformed, and then deformation ended and the upper, undeformed units were deposited. Although any amount of rock emplacement and rock deformation can occur, and they can occur any number of times, these concepts provide a guide to understanding the geological history of an area.

Methods of Geology

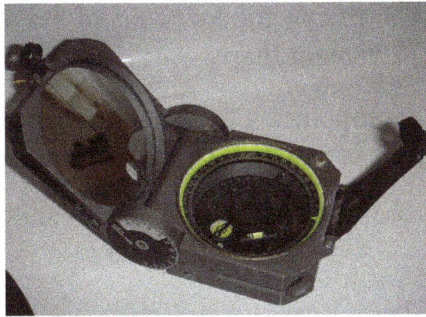

A standard Brunton Pocket Transit, commonly used
by geologists for mapping and surveying.

Geologists use a number of field, laboratory, and numerical modeling methods to decipher Earth history and to understand the processes that occur on and inside the Earth. In typical geological investigations, geologists use primary information related to petrology (the study of rocks), stratigraphy (the study of sedimentary layers), and structural geology (the study of positions of rock units and their deformation). In many cases, geologists also study modern soils, rivers, landscapes, and glaciers; investigate past and current life and biogeochemical pathways, and use geophysical methods to investigate the subsurface. Sub-specialities of geology may distinguish endogenous and exogenous geology.

Field Methods

A typical USGS field mapping camp.

Today, handheld computers with GPS and
geographic information systems software are often
used in geological field work (digital geologic mapping).

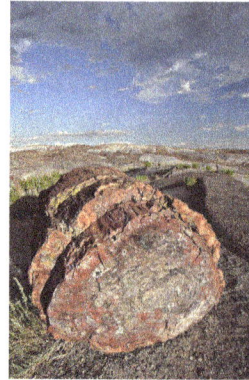

A petrified log in Petrified Forest
National Park, Arizona, U.S.A.

Geological field work varies depending on the task at hand. Typical fieldwork could consist of:

- Geological mapping:

  ○ Structural mapping: identifying the locations of major rock units and the faults and
    folds that led to their placement there,

  ○ Stratigraphic mapping: pinpointing the locations of sedimentary facies (lithofacies and
    biofacies) or the mapping of isopachs of equal thickness of sedimentary rock,

  ○ Surficial mapping: recording the locations of soils and surficial deposits.

- Surveying of topographic features:

  ○ Compilation of topographic maps,

  ○ Work to understand change across landscapes, including:

    ▪ Patterns of erosion and deposition,

    ▪ River-channel change through migration and avulsion,

    ▪ Hillslope processes.

- Subsurface mapping through geophysical methods:

  ○ These methods include:

    ▪ Shallow seismic surveys,

    ▪ Ground-penetrating radar,

    ▪ Aeromagnetic surveys,

    ▪ Electrical resistivity tomography.

  ○ They aid in:

    ▪ Hydrocarbon exploration,

- Finding groundwater,
- Locating buried archaeological artifacts.

- High-resolution stratigraphy:
  - Measuring and describing stratigraphic sections on the surface,
  - Well drilling and logging.

- Biogeochemistry and geomicrobiology:
  - Collecting samples to:
    - Determine biochemical pathways,
    - Identify new species of organisms,
    - Identify new chemical compounds.
  - And to use these discoveries to:
    - Understand early life on earth and how it functioned and metabolized,
    - Find important compounds for use in pharmaceuticals.

- Paleontology: excavation of fossil material:
  - For research into past life and evolution,
  - For museums and education.

- Collection of samples for geochronology and thermochronology,

- Glaciology: measurement of characteristics of glaciers and their motion.

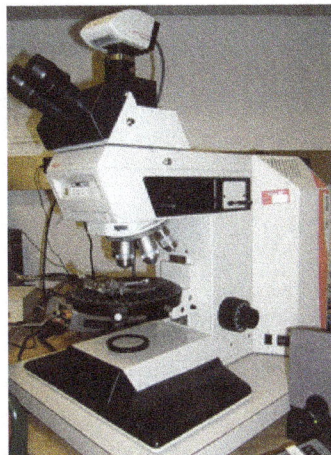

A petrographic microscope – an optical microscope fitted with cross-polarizing lenses, a conoscopic lens, and compensators (plates of anisotropic materials; gypsum plates and quartz wedges are common), for crystallographic analysis.

Petrology

Folded rock strata.

In addition to identifying rocks in the field (lithology), petrologists identify rock samples in the laboratory. Two of the primary methods for identifying rocks in the laboratory are through optical microscopy and by using an electron microprobe. In an optical mineralogy analysis, petrologists analyze thin sections of rock samples using a petrographic microscope, where the minerals can be identified through their different properties in plane-polarized and cross-polarized light, including their birefringence, pleochroism, twinning, and interference properties with a conoscopic lens. In the electron microprobe, individual locations are analyzed for their exact chemical compositions and variation in composition within individual crystals. Stable and radioactive isotope studies provide insight into the geochemical evolution of rock units.

Petrologists can also use fluid inclusion data and perform high temperature and pressure physical experiments to understand the temperatures and pressures at which different mineral phases appear, and how they change through igneous and metamorphic processes. This research can be extrapolated to the field to understand metamorphic processes and the conditions of crystallization of igneous rocks. This work can also help to explain processes that occur within the Earth, such as subduction and magma chamber evolution.

Structural Geology

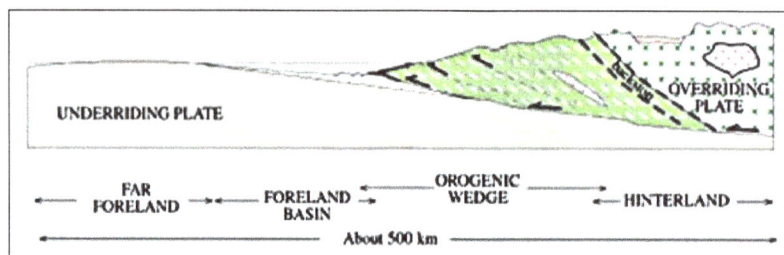

A diagram of an orogenic wedge.

The wedge grows through faulting in the interior and along the main basal fault, called the décollement. It builds its shape into a critical taper, in which the angles within the wedge remain the same as failures inside the material balance failures along the décollement. It is analogous to a bulldozer pushing a pile of dirt, where the bulldozer is the overriding plate.

Structural geologists use microscopic analysis of oriented thin sections of geologic samples to observe the fabric within the rocks, which gives information about strain within the crystalline structure of the rocks. They also plot and combine measurements of geological structures to better understand the orientations of faults and folds to reconstruct the history of rock deformation in the area. In addition, they perform analog and numerical experiments of rock deformation in large and small settings.

The analysis of structures is often accomplished by plotting the orientations of various features onto stereonets. A stereonet is a stereographic projection of a sphere onto a plane, in which planes are projected as lines and lines are projected as points. These can be used to find the locations of fold axes, relationships between faults, and relationships between other geologic structures.

Among the most well-known experiments in structural geology are those involving orogenic wedges, which are zones in which mountains are built along convergent tectonic plate boundaries. In the analog versions of these experiments, horizontal layers of sand are pulled along a lower surface into a back stop, which results in realistic-looking patterns of faulting and the growth of a critically tapered (all angles remain the same) orogenic wedge. Numerical models work in the same way as these analog models, though they are often more sophisticated and can include patterns of erosion and uplift in the mountain belt. This helps to show the relationship between erosion and the shape of a mountain range. These studies can also give useful information about pathways for metamorphism through pressure, temperature, space, and time.

Stratigraphy

Different colours show the different minerals composing the mount
Ritagli di Lecca seen from Fondachelli-Fantina, Sicily.

In the laboratory, stratigraphers analyze samples of stratigraphic sections that can be returned from the field, such as those from drill cores. Stratigraphers also analyze data from geophysical surveys that show the locations of stratigraphic units in the subsurface. Geophysical data and well logs can be combined to produce a better view of the subsurface, and stratigraphers often use computer programs to do this in three dimensions. Stratigraphers can then use these data to reconstruct ancient processes occurring on the surface of the Earth, interpret past environments, and locate areas for water, coal, and hydrocarbon extraction.

In the laboratory, biostratigraphers analyze rock samples from outcrop and drill cores for the fossils found in them. These fossils help scientists to date the core and to understand the depositional environment in which the rock units formed. Geochronologists precisely date rocks within the stratigraphic section to provide better absolute bounds on the timing and rates of deposition. Magnetic stratigraphers look for signs of magnetic reversals in igneous rock units within the drill cores. Other scientists perform stable-isotope studies on the rocks to gain information about past climate.

Planetary Geology

Surface of Mars as photographed by the
Viking 2 lander.

With the advent of space exploration in the twentieth century, geologists have begun to look at other planetary bodies in the same ways that have been developed to study the Earth. This new field of study is called planetary geology (sometimes known as astrogeology) and relies on known geologic principles to study other bodies of the solar system.

"Geology" is often used in conjunction with the names of other planetary bodies when describing their composition and internal processes: examples are "the geology of Mars" and "Lunar geology". Specialised terms such as selenology (studies of the Moon), areology (of Mars), etc., are also in use.

Although planetary geologists are interested in studying all aspects of other planets, a significant focus is to search for evidence of past or present life on other worlds. This has led to many missions whose primary or ancillary purpose is to examine planetary bodies for evidence of life. One of these is the Phoenix lander, which analyzed Martian polar soil for water, chemical, and mineralogical constituents related to biological processes.

Applied Geology

Economic Geology

Man panning for gold on the Mokelumne.

Economic geology is a branch of geology that deals with aspects of economic minerals that human-kind uses to fulfill various needs. Economic minerals are those extracted profitably for various practical uses. Economic geologists help locate and manage the Earth's natural resources, such as petroleum and coal, as well as mineral resources, which include metals such as iron, copper, and uranium.

Mining Geology

Mining geology consists of the extractions of mineral resources from the Earth. Some resources of economic interests include gemstones, metals such as gold and copper, and many minerals such as asbestos, perlite, mica, phosphates, zeolites, clay, pumice, quartz, and silica, as well as elements such as sulfur, chlorine, and helium.

Petroleum Geology

Mud log in process, a common way to study the lithology when drilling oil wells.

Petroleum geologists study the locations of the subsurface of the Earth that can contain extractable hydrocarbons, especially petroleum and natural gas. Because many of these reservoirs are found in sedimentary basins, they study the formation of these basins, as well as their sedimentary and tectonic evolution and the present-day positions of the rock units.

Engineering Geology

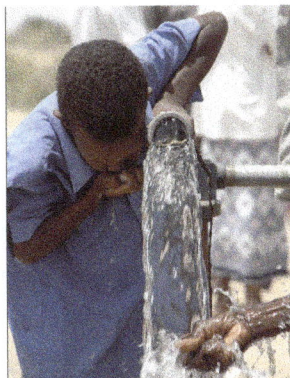

A child drinks water from a well built as part of a hydrogeological humanitarian project in Shant Abak, Kenya.

Engineering geology is the application of the geologic principles to engineering practice for the purpose of assuring that the geologic factors affecting the location, design, construction, operation, and maintenance of engineering works are properly addressed.

In the field of civil engineering, geological principles and analyses are used in order to ascertain the mechanical principles of the material on which structures are built. This allows tunnels to be built without collapsing, bridges and skyscrapers to be built with sturdy foundations, and buildings to be built that will not settle in clay and mud.

Hydrology and Environmental Issues

Geology and geologic principles can be applied to various environmental problems such as stream restoration, the restoration of brownfields, and the understanding of the interaction between natural habitat and the geologic environment. Groundwater hydrology, or hydrogeology, is used to locate groundwater, which can often provide a ready supply of uncontaminated water and is especially important in arid regions, and to monitor the spread of contaminants in groundwater wells.

Geologists also obtain data through stratigraphy, boreholes, core samples, and ice cores. Ice cores and sediment cores are used to for paleoclimate reconstructions, which tell geologists about past and present temperature, precipitation, and sea level across the globe. These datasets are our primary source of information on global climate change outside of instrumental data.

Natural Hazards

Geologists and geophysicists study natural hazards in order to enact safe building codes and warning systems that are used to prevent loss of property and life.

Geomorphology

Geomorphology is "a discussion on Earth's form". Hence, it is the study of various features that are found on the Earth, such as mountains, hills, plains, rivers, moraines, cirques, sand dunes, beaches, spits, etc., that are created by various agents such as rivers, glaciers, wind, ocean, etc. Since the fourth century BC, many people have studied the formation of the Earth by relating to various observations in the field. Ancient Greeks and Romans such as Aristotle, Strabo, Herodotus, Xenophanes, and many others discussed about the origin of the valleys, formation of deltas, presence of seashells on mountains, etc. After observing the seashells on the top of the mountains, Xenophanes speculated that the surface of the Earth must have risen and fallen from time to time, thus creating river valleys and mountains. After observing seashells on mountain top and vast expanses of sand, Aristotle suggested that the areas which are dryland now must be covered by sea in the past and those areas where sea is present now must have been dryland once. Hence, he proposed that land and sea change places. Traditionally, the history of the development of landscape was carried out by mapping the sedimentary and morphological features. For understanding the evolution of landscape, the golden rule, "the present is the key to the past," has been followed. This rule assumes that the processes that are visible in action today must have occurred in the past also, which can be used to infer the reasons for formation of the landscape in the past. Hence, the past formation was mainly dependent on the relative information and aging method.

However, the word "geomorphology" was first coined and used between the 1870s and 1880s to describe the morphology of the surface of the Earth. But it was popularized by William Morris

Davis who proposed the "geographical cycle" also known as "Davis cycle". He proposed that the development of landscapes occurs as due to alternate action of uplift and denudation. He assumed that uplift occurs quickly and then the uplifted land mass erodes gradually to form the topography of the region. He hypothesized that upliftment is a quick action, whereas denudation is a time-taking process. Thus, creating high mountains and deep valleys showcases youth, mature, and old age stages of landform development. Though Davis cycle is considered as a classic work, but his hypothesis lacks a basic understanding that both upliftment and denudation occur simultaneously. Both of these phenomena go on hand in hand and are not necessarily alternate. Hence, nearly 35 years later, Walther Penck proposed a variation of "Davis model," where he showed that the interaction of both uplift and denudation occurs simultaneously. He suggested that due to simultaneous actions, the slopes will be developed in three main forms. First, a convex slope where the upliftment rate is higher than denudation rate; next, a steady-state or stationary stage where both the rates are nearly equal, hence creating straight slope; and finally concave slopes when the rate of upliftment is lesser than the rate of denudation. Thus, over a period of time, various aspects of landforms have been studied by geomorphologists. Some geomorphologists have studied the process of formation of these landforms, while some have studied its origin and history, and others have analyzed various forms of landforms for their quantitativeness. Hence, in a nutshell, modern geomorphologists focus mainly on three aspects of landforms: form, process, and history. The form and process studies are commonly termed as functional geomorphology, while the last one as historical geomorphology. The study of various processes that are responsible for the formation of a landscape falls in the purview of functional geomorphology.

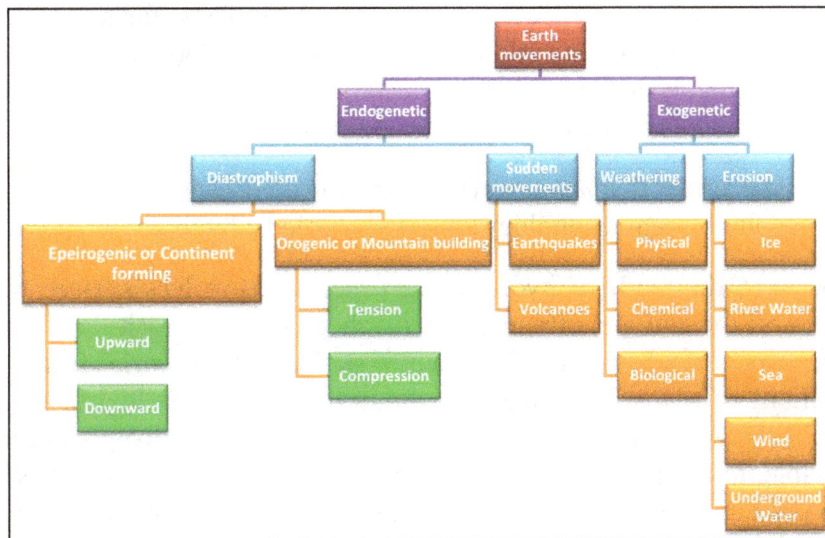

Endogenic and exogenic processes.

All these landforms that are visible on the Earth vary in size from microscale such as potholes, flutes, ripples, etc. to mega-scale features such as mountain ranges, river basins, etc. Hence, the time required to form these features also varies from tens of years to millions of years. It has also been observed that certain features are native to certain climatic zones; hence, development of climatic zones such as arid, tropical, etc. plays a critical role in formation and evolution of these geomorphic features. For example, the landforms observed in higher latitudes show signature of glaciation and deglaciation cycle, which is indicative of quaternary climatic environment, whereas in other parts of the world, such as Grand Canyon of Colorado River Valley in the United States have preserved the signature of various activities, that occurred hundreds of millions of years ago, in its various landforms. Most of the landforms are formed and deformed due to the two processes, namely, endogenic that occurs within the Earth's crust such as convective heat cycles, rising plume, and magma chambers and exogenic processes that shape the features on the surface of the Earth with the help of various agents of weathering like water, wind, glaciers, seas, etc.

A lot of works have been carried out in the field of functional and historical geomorphology. Now, many other fields or kinds of geomorphology have been studied such as tectonic geomorphology, submarine geomorphology, planetary geomorphology, climatic geomorphology, and modeling geomorphology. Interaction of tectonic forces and geomorphic processes deform the Earth's crust regularly, and this led to development of tectonic geomorphology. It uses the techniques and data from other fields of geology mainly structural, geochemistry, geochronology in conjunction with geomorphology, and climate change. As name suggests, submarine geomorphology focuses on the origin, formation, and development of submarine landforms developed in both shallow and deep marine environments. Planetary geomorphology deals with the application of the understanding of the formation of landforms on the Earth to extraterrestrial objects such as moon, planets, exoplanets, etc. This is comparatively the latest branch and is developing very fast. Geomorphic studies of Venus, Mars, Jupiter, Titan, and other planets are a hot cake these days. Climate plays a critical role in developing various landform natives to each climatic zone such as arid, tropical, temperate, etc. This understanding is the basis for the development of climatic geomorphology as a stream. The effect of climatic phenomena along with tectonic activities places a new cross stream of geomorphology known as climato-tectonic geomorphology. These days, inter- and multidisciplinary approaches have been used in various fields of science, and geomorphology is one of them where cross breeding is highly evident. Till now, various branches and offshoots of geomorphology have been developed, and lots of researches have been carried out in those interdisciplinary areas.

Among all the exogenic agents that are at work to form the landscape, water is the most promising and effective. Hence, fluvial geomorphology has been studied a lot in details.

Hydro-geomorphology, the study of hydrological processes, involves surface runoff, baseflow, stream discharge, and the soil and streambed erosion processes, which continuously chisel the geomorphological profile of a basin. The life span of such processes varies from few hundreds of years to even millions of years. Apart from the quantification of the hydrological processes, as well as the soil and streambed erosion processes, the continuous hydro-geomorphologic modeling provides valuable information for the future trend of these physical processes. A wide variety of integrated models that continuously simulate the runoff, soil erosion, and sediment transport processes are available.

Anthropogenic activities have significantly affected the fluvial geomorphological regimes within a very short time span. From construction of dams which increases the sedimentation in the reservoir, thus changing the riverbed profile to the deforestation and urbanization which is enhancing the erosion rates in the river catchment, anthropogenic activities have left its imprints in the natural phenomenon. Similar is the case in St. Lawrence Lowlands of Quebec region of the Canadian Shield where the construction of dams has led to increase in bank-full width, thus decreasing the channel sinuosity and changing the fluvial regimes. Further changes in land-use pattern have also led to higher erosion and sedimentation. Clearing of forests for agricultural practices has led to deforestation, and later afforestation in that region (agricultural areas) due to decline in agricultural work force has impacted the morphological evolution of channels in Quebec region of Canada.

Higher rates of erosion are observed when the weathering agent is water. And, considering the huge expanses of oceans and the erosion that occurs at the shores takes the first place. This effect is clearly visible in shoreline change and sea-level rise. Most of the populated cities all around the world are situated near the coasts; thus, majority of the population of the world lives within few kilometers of the coast. Thus, a proper coastal land management is required to cater the needs of the ever-increasing population. Shoreline change (cliff erosion) has been studied using predictive models which are based on historical records and the geomorphological data of a certain region. Current historical extrapolation models use historical recession data, but different environments with the same historical values can produce identical annual retreat characteristics despite the potential responses to a changing environment being unequal.

With the advent of satellite technology, it has been absolutely easy to study the surface of the Earth from satellite data. When it comes to identifying various landforms and describing the physical appearances, satellite images or aerial photos come very handy. However, this approach is more qualitative than quantitative and is defined as morphography, where the external shapes are described without giving information about the way of creation of those features. Various methods are used to define the origin of features and the mechanism of development of these features. This comes under morphogenesis, while morphochronology deals with the estimation of age of the forms in the absolute as well as relative terms. Finally, the quantitative estimation carried out by measurements of the geometric features of the landforms is known as morphometry. There are various morphometric parameters and morphometric indices being used in geomorphometry to define the landform analysis and classification.

References

- Earth-sciences, science: britannica.com, Retrieved 15 March, 2019

- Geography: worldatlas.com, Retrieved 23 July, 2019

- Introductory-chapter-geomorphology, hydro-geomorphology-models-and-trends: intechopen.com, Retrieved 16 June, 2019

- Taylor, Stuart Ross (29 July 2004). "Why can't planets be like stars?". Nature. 430 (6999): 509. Bibcode:-2004Natur.430..509T. doi:10.1038/430509a. PMID 15282586

- Das, Braja M. (2006). Principles of geotechnical engineering. England: Thomson Learning. ISBN 978-0-534-55144-5

Permissions

All chapters in this book are published with permission under the Creative Commons Attribution Share Alike License or equivalent. Every chapter published in this book has been scrutinized by our experts. Their significance has been extensively debated. The topics covered herein carry significant information for a comprehensive understanding. They may even be implemented as practical applications or may be referred to as a beginning point for further studies.

We would like to thank the editorial team for lending their expertise to make the book truly unique. They have played a crucial role in the development of this book. Without their invaluable contributions this book wouldn't have been possible. They have made vital efforts to compile up to date information on the varied aspects of this subject to make this book a valuable addition to the collection of many professionals and students.

This book was conceptualized with the vision of imparting up-to-date and integrated information in this field. To ensure the same, a matchless editorial board was set up. Every individual on the board went through rigorous rounds of assessment to prove their worth. After which they invested a large part of their time researching and compiling the most relevant data for our readers.

The editorial board has been involved in producing this book since its inception. They have spent rigorous hours researching and exploring the diverse topics which have resulted in the successful publishing of this book. They have passed on their knowledge of decades through this book. To expedite this challenging task, the publisher supported the team at every step. A small team of assistant editors was also appointed to further simplify the editing procedure and attain best results for the readers.

Apart from the editorial board, the designing team has also invested a significant amount of their time in understanding the subject and creating the most relevant covers. They scrutinized every image to scout for the most suitable representation of the subject and create an appropriate cover for the book.

The publishing team has been an ardent support to the editorial, designing and production team. Their endless efforts to recruit the best for this project, has resulted in the accomplishment of this book. They are a veteran in the field of academics and their pool of knowledge is as vast as their experience in printing. Their expertise and guidance has proved useful at every step. Their uncompromising quality standards have made this book an exceptional effort. Their encouragement from time to time has been an inspiration for everyone.

The publisher and the editorial board hope that this book will prove to be a valuable piece of knowledge for students, practitioners and scholars across the globe.

Index

www.ingramcontent.com/pod-product-compliance
Lightning Source LLC
Chambersburg PA
CBHW082027190326
41458CB00010B/3293